The Last Chance Energy Book

Also by Owen Phillips

The Dynamics of the Upper Ocean
The Heart of the Earth

The Last Chance ENERGY Book

Owen Phillips

The Johns Hopkins University Press
Baltimore and London

Copyright © 1979 by Owen Phillips

All rights reserved. No part of this book may be
reproduced or transmitted in any form or by any means,
electronic or mechanical, including photocopying,
recording, xerography, or any information storage and
retrieval system, without permission in writing
from the publisher.

Manufactured in the United States of America

The Johns Hopkins University Press, Baltimore, Maryland 21218
The Johns Hopkins Press Ltd., London

Originally published, 1979
Second printing, 1979

Library of Congress Catalog Card Number 78–20511
ISBN 0–8018–2189–4

Library of Congress Cataloging in Publication data
will be found on the last printed page of this book.

Illustrations by Charles West

*To Lynette, Christopher, Bronwyn, and Michael,
who must live through this
unwanted revolution*

Contents

Preface
ix

**1
The Unwanted Revolution**
1

**2
In the Time Before Morning**
12

**3
The Profligate Age**
22

**4
Growth and Decay:
The Inexorable Cycles**
27

5
Megapower and Micropower
52

6
Personal Choices:
A Survival Guide for the Coming Crunch
63

7
Finding the Missing Megapower:
Immediate Alternatives
79

8
Finding the Missing Megapower:
New Sources
102

9
The Last Chance
127

Index
135

Preface

This book seeks to explain simply and clearly the frightening magnitude of the energy crunch, how we know what the real facts are, why the problem will become worse before it gets better, and what the sensible options are on a personal and national level. It is intended to help the individual reader cope with the unwanted revolution that is being forced upon him and, frankly, to stimulate action on a national level before the time runs out.

The United States has faced many "crises" in the recent past, and somehow we have muddled through; things have usually turned out to be not really as bad as the doomsday prophets foretold. It would be pleasant to believe that the same will again be true, but alas, wishing will not make it so. The energy problem has a vastness and an inevitability that may leave us with a sense of individual helplessness. It involves technical, economic, and societal issues which we may feel to be beyond our individual interest or our ability to cope. It is threatening to the fabric of our society, to the lives that we know. Yet we, the people, must live through the problem and we, the people, must ultimately pay for its solution.

This is not a technical book, nor, I think, should it be. We do not need to immerse ourselves in technicalities to appreciate the true dimensions of the energy problem and to place in some perspective the possibilities there are for solution. Yet, while avoiding technicalities, I have made every effort to ensure that the numbers quoted and the facts given are correct. "Estimates" are often made on the flimsiest of bases; too often they are used to becloud the issues. The facts speak for themselves; they must not be hidden and they certainly need no exaggeration.

This is not a doomsday book either. The facts will be found to be alarming and, with inaction, the prospects will be worse. Nevertheless, I have an unbounded faith in the determination, ability, and ingenuity of the American people who, when faced with a problem even of this magnitude, will find solutions with which we can live. The qualities that we need desperately are there, certainly unfocused, perhaps still latent. If this book plays some small part in bringing them together, it will have served its purpose. It is dedicated specifically to my children whose quality of life depends on the outcome, and more generally to all young people for whom this country is held in trust.

I am grateful for the assistance of many people in the preparation of this book, but particularly to Dr. William Avery and Mr. Charles Lamb for drawing attention to information that I would otherwise have missed, Ms. Joanne Koutz for her help with the research and preparation of the manuscript and Dr. Donald Zinn for his many constructive comments. They have contributed perhaps more than they know.

The Last Chance Energy Book

1
The Unwanted Revolution

The theme of this book is simple: We are being forced into a revolution, a series of profound changes that will permeate almost every aspect of our daily living. It is not a political revolution, inspired by a determined cell of radicals of the right or the left, although it could become one. It is not merely a technological revolution or a social one, but it does encompass both. Yet it is a revolution perhaps more dangerous and more complex than any that had to be faced in the past.

Change, in itself, is not unusual. We live with change, we expect it. The development of the automobile as a practical, relatively reliable means of transport has precipitated one of the many revolutions within this century. We have virtually forgotten that in 1900, a journey of one hundred miles was a substantial and arduous undertaking, unless it could be made by rail or by sea. We took the automobile to our hearts and it transformed for better or worse not only our conception of distance but the structure and quality of our cities and the styles of our lives. The whole avalanche of technical innovation during the last hundred and fifty years, based on an abundance of cheap energy, has created a greater change in the material standard of living during that relatively short time than had taken place over the previous two millennia. For the most part, we welcomed the change; at any rate, we accepted it in spite of the accompanying alteration in the quality of the environment in which this material life was pursued.

But this new revolution is different in several respects: it is not welcome, it is threatening, and it is being forced upon us. It is inevitable, it is inexorable. Yet its course can be molded by choices, individual choices and collective choices in which, like it or not, everyone will be obliged to play his part. The force behind this revolution is, of course, the problem of energy. Our experiences to date and our responses to them represent only the first skirmishes in what could become a protracted and debilitating war. It can, if we let it, impose crushing constraints on the society that we know; it can impoverish us individually and collectively, and conceivably even reduce us to the condition in which the mere sustenance of life is the main business of living. However, such an outcome is not inevitable; although the revolution will come, it can be guided. To shape it for the betterment of our society will require individual imagination, initiative, and daring, together with a collective determination based on an understanding of the problems and the sensible options that are available for their solution. The choices may sometimes be difficult, involving the weighing of opposing sets of values, but they must be made. Even the choice of doing nothing sets us along a certain road, but one that we may not wish to follow and from which it may be difficult to turn.

As a concerned citizen, you may well ask: "What is the truth about this energy problem? Is it as serious as some people say? Why should I believe them and how do they know? True, during the oil embargo we had to line up for our gasoline, but after a time the price went up and the lines melted away. How do I know that we are not simply being manipulated, that the whole thing will not disappear when natural gas is de-regulated and some semblance of tranquility is restored to the Middle East? Even if there is a problem, what can I do about it? How is it going to affect me and my children? What choices are there for me to make?" These important questions deserve honest answers. The oil companies may give one set of responses, the utility companies and automobile manufacturers others, each honest enough but each crafted to suit particular interests and expectations. Solar and geothermal energy each have their devotees and advocates; some people would have us build windmills or devices to extract energy from the waves of the sea. There is no shortage of experts on energy who are ready with answers—they appear like fireflies on a summer night. Each has his own vision and marches to the beat of his own drummer. But what is the individual citizen to make of the clamor of such a multitude of experts? He has learned to be skeptical of even authoritative pronouncements. He needs to be convinced, not by the word of an expert, but by seeing for himself. What legal testimony would have convinced us of the ramifications of Watergate had we not heard or read for ourselves the contents of those infamous tapes? We need to see not only what the facts are but the basis for them, what is real and what is rhetoric, what is sensible and what is fantasy, what are the choices and what are the sensible alternatives. It is we who must make the choices, either personally or through our legislative representatives and, in the final analysis, it is we who must bear the costs.

We do not need to immerse ourselves in the technicalities of BTU's, kilowatt-hours, or conversion efficiencies—leave that to the professionals. There are a number of very serious and rather technical books and reports on the energy problem;[1] important as they are, they are rather inaccessible to most of us. We need perspective rather than technical detail, a basis for judgment rather than a specific blueprint for solution.

Perhaps the one point on which most experts agree is that there is indeed an energy problem. It is most acute with respect to oil: we are

[1] One very good one is by John M. Fowler, *Energy and the Environment* (New York: McGraw-Hill, 1975).

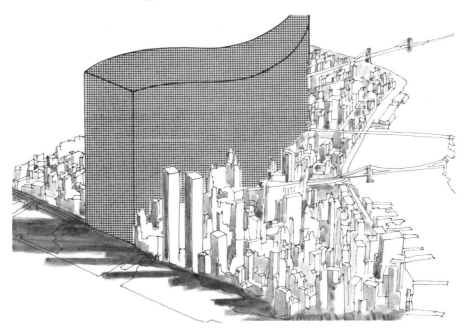

heavily dependent on oil in almost every phase of our life, the world is running very quickly out of oil, and once it is gone, there is no more. *In the United States we have already discovered and have already consumed more than half the producible oil that was created by nature over hundreds of millions of years.* We have done this in fifty years, a brief interval since this country was founded, and a mere blip of time in the recorded history of mankind. In fifty years we have used over half of all there ever was, and at the rate we're going, we'll finish off the rest very quickly.

Some simple figures will illustrate the magnitude of the problem. The United States consumes about six billion barrels of oil per year. This is an enormous quantity, almost impossible to visualize. Perhaps one could calculate that if the six billion barrels were stacked into a wall thirty feet thick and fifty miles long, it would reach more than half a mile high, dwarfing the highest buildings in New York City. But this image makes little sense. Fortunately, it is not necessary to try to imagine six billion barrels of oil—let us just remember the number *six*.

The United States used to produce most of the oil that we consumed; we now produce about three billion barrels and import the rest. Incidentally, we import it at a great expense to our economy—at about $13 a barrel, the bill comes out to some $40 billion per year and rising, just for oil. The immensity of the bill is one of the principal causes of the deficit in our balance of payments, the declining value of the dollar in money markets abroad, and the stubbornly high rates of inflation. Quite aside from these questions, the known reserves of oil in the United States are approximately thirty-one billion barrels;[2] we presently extract three billion per year, and if we kept going at this rate the reserves would last about ten years. Not the far and indefinite future, just ten years.

Known reserves represent oil in the ground which is extractable and whose existence has been proven by drilling. What about the possibility of further new finds in Alaska? What about the off-shore fields that we are told may exist along the Atlantic continental shelf? The off-shore fields are potentially very important, though not proven. To rely on these for our future energy needs is rather like relying on a rally of the stock market to repay our creditors. How much the stock market is going to rally, if at all, is not known any better than how much oil is really there. We may be lucky, we may not. The techniques for *finding geological structures* that may contain oil deep inside the earth are quite sophisticated, but even if such structures are found, they may or they may not contain oil. The methods used to estimate the *amount of oil* that might be present in a particular area are, by contrast, astonishingly primitive, often amounting to little more than a guess. Nevertheless, if one does make an optimistic estimate, an upper-limit guess, for the amount of oil that might be along the entire Atlantic continental shelf, the total amount recoverable turns out to be at most *six billion barrels*. In terms of present consumption, this is just one year's supply. For Alaska, it is perhaps another twenty billion—three years' supply. Even allowing for other large finds not yet contemplated—and there cannot be too many more of these—we might add a few more years, maybe ten at the outside, but the conclusion is inescapable. The United States is going to run out of oil very soon.

[2] *Oil and Gas Journal*, 18 April 1977, p. 19.

Domestic production is already declining. To make up for the difference, we are being forced to import oil at whatever the cost. At the time of the oil embargo the price tripled; it has escalated steadily since then, and the brutal fact is that it will continue to do so. There is simply no prospect of relief as we sink further into the position of having to import more and more of what, even on a worldwide scale, is a diminishing resource. The real point is that even all *the total oil resources of the whole world are finite and they are nonrenewable.* What is true for the United States is true, on a very slightly different time scale, for the world as a whole. Unlike a forest that can be replanted, an oil field, once depleted, is finished. Again, if we hopefully allow for discoveries not yet made and do our optimistic calculations, it turns out that the world oil resources will support present world consumption for only about thirty years or so.

For natural gas, it will be found that the prospects are depressingly similar. Both oil and natural gas are short lived resources that can be exhausted within the lifetime of those who are young adults today. The implications of this prospect are immense. At present, we derive 75% of our energy from these sources, not only to power our automobiles and heat our homes, but to extract other raw materials from the ground, to manufacture furniture and fertilizers, cement for housing, steel, aluminum and plastics. Energy derived from oil and natural gas is used in the brewing of beer, in building the machines that make other machines, in the basement workshop, and in producing our newspapers. Three-quarters of the energy sources underlying all these things will shortly be exhausted—to do without the energy would wrench the very fabric of our society in ways that are almost unimaginable; to replace it will require determination that we have not begun to muster. This is only a hint of the extent of the revolution that is being forced upon us.

These statements are alarming, but do they represent a cry of "Wolf, wolf" when no wolf is there? A reasonable person may well ask whether in fact such dire predictions are justified. How much can one rely on the calculations that underlie them? Some people have asserted that there is no problem—Mr. Henry Ford II is reported to have said "Nobody knows how much oil is left in the ground. And the chances are that we will never find out because we will never get to the bottom of the barrel."³ We would all prefer to believe him. However, as we will see, we do know approximately how much oil is left in the ground, both on a national and a global scale, even before we get to the bottom of the barrel; the way in which the facts are established is simple and the logic inescapable. We will see for ourselves what the present situation is and how we know; we can then make our own judgments. There are choices to be made and we will see what some of the sensible alternatives might be.

Suppose we can be convinced that there is a problem? Our first response, rightly, should be to practice conservation and the ever-increasing cost of these fuels is bound to provide a powerful incentive. Conservation is certainly very important but it is only a small part of the answer. Oil conservation would be the answer only if there were a continuous but limited supply. If the water for my house comes from a permanent spring, then I must limit my consumption of water to the amount that flows from the spring. If I practice conservation of water, there is no problem and I remain in good shape. But if my water comes from a tank and there is no new supply, I can cut my consumption by ten percent or fifty percent or whatever to make it last longer, but I am going to run out of water in the end. Unfortunately, this latter situation is the one that faces us with oil.

Conservation on an individual scale will save us money. Nationally, if conservation is our only strategy, it can give us a reprieve for a couple of years, but that is about all.

In our personal use of energy, we are perhaps most conscious (or guilty?) about the amount of gasoline that is pumped into our automobiles. What will we do with them? Few of us would feel inclined to go about on a horse, although perhaps we might use a bicycle if we do not have to go too far. Nor will we have cars driven directly by windmills or solar panels—the power that these devices can provide is simply insufficient. It seems much more likely that we

³ *New York Times*, 1 August 1976.

will be forced by the increasing price of fuel or perhaps by rationing to limit our use of automobiles at least until some alternative fuels can be produced in sufficient quantity. Sensible alternatives are known, but technologically we are a long way from where we must be if they are to be available in time.

In the search for new energy sources, the sun is an obvious candidate. It is generally believed that solar energy is surely the long-term answer, and this belief is almost certainly justified. In the United States, we have been somewhat behind other countries in developing solar power. Even a casual visitor to Western Australia would notice that many of the new houses there have solar panels built into their roofs. It is quite routine, not at all exceptional. That part of the world has always been energy poor—there is no oil, not much coal (the deposits that do exist being of indifferent quality), and little hydroelectric power. On the other hand, the climate is generally sunny with about a month of cloudy weather per year so that solar panels, together with a small back-up system, provide plenty of hot water for heating and for household use. Why cannot we do the same thing in this country on a larger scale?

No doubt we shall; this is one of the personal options open to us. It takes no great foresight to anticipate that domestic solar panels will become increasingly more economical over the long run as the technology for their manufacture improves and as the prices of alternative fuels rise. They will make a great difference to our personal budgets and will help to reduce the national demand for oil and natural gas. But solar panels are not for everyone. What about all the people who live in apartment houses, hotels, and condominiums? Or in areas of the country where the sun shines only intermittently in winter, precisely the time when the heat is needed most? It is probably an over-statement of the case, but not an outrageous one, to assert that domestic solar collectors will provide for the energy problem the same kind of contribution that backyard vegetable plots provide for our food supply—sufficient for a relatively few fortunate people and a valuable supplement for others, but that's all.

Solar energy may indeed be the long-term answer, but the use of domestic collectors is not the only way to capture it. Another option would be to cover the deserts with solar collectors, using the heat to generate electric power. But there are problems. Do we really want to cover the deserts with solar collectors? New Englanders may not mind, but the people who live in Arizona may be less than enthusiastic. The ecology of the desert is a fragile thing and the impact of large farms of solar collectors may be difficult to assess. Are we prepared for the enormous capital costs? Solar collectors work very well when the sun is shining, but they are not very useful at night. Energy would need to be stored in very large amounts to carry us through for nighttime use or during extended cloudy periods; either this or a back-up system of large capacity that is used only intermittently and is consequently expensive. Solar power from the deserts may well be part of the solution, but we should not assume that it is the whole solution.

Nevertheless, there are other ways of gathering energy from the sun, less well-known, that may offer even greater promise. Most of the surface of the earth, about two-thirds, is covered by sea, so that two-thirds of the solar energy ends up in the ocean. The ocean is the biggest collector of solar energy that we could conceive. One way of recovering a small fraction of this energy for our use is to take advantage of temperature differences in the ocean to generate electric power on a large scale. At first acquaintance, this idea seems ridiculous. A number of proposals that have been put forward to solve the energy problem have been ridiculous, but this particular one is not. It was in fact demonstrated as long ago as 1930 by the French engineer Claude, and the process worked. To be sure, it did not work very well—his arrangement was very crude by today's standards—but it did work. The first steam locomotive did not work very efficiently either.

There are still other options, less esoteric. Nuclear power plants are already contributing substantially to our electrical energy needs, but their contribution must be kept in perspective. Nuclear power presently satisfies about 2% of our national appetite for energy; oil and natural gas about 75%. Nuclear power generation is very costly and would need to be expanded enormously if it is to make up for a significant fraction of the loss of fossil fuels. Already there are serious problems with the disposal of radioactive wastes. The overall safety record of nuclear power plants has been very good; safeguards have been stringent. Nevertheless, there is the apprehension of environmental problems of a possibly insidious and subtle kind so that popular enthusiasm for such plants has generally been mixed. Quite aside from that, the uranium isotope U235 used in present power plant reactors is itself in limited supply. Uranium is not a common metal in the first place and U235 constitutes less than 1% of the natural metal. In some ways it is the same story as with oil and gas, though there are some interesting differences to it.

No doubt, nuclear power plants will generate an increasing proportion of our electrical energy, though not as much as was once envisioned, in spite of the heavy capital expense involved in construction, in spite of the large amounts of waste heat that these plants discharge into the environment, and in spite of the problems of disposal of radioactive wastes. But one may well suspect that these problems will, within a limited time, become intolerable and that nuclear power is unlikely to be a long-term answer.

These and other options we must explore. There is no single answer, no panacea that can stave off the revolution. We must simply cope with it as it develops.

The "alarming statements and unsupported assertions" made earlier in this chapter are unfortunately true, but before they can be believed, we must see why they are true. If they be granted, the exercise of energy conservation and the development of alternative energy sources are essential on a personal and national level, but they will take time. Until then, we will continue to rely on oil and natural gas to a considerable extent, even as their price continues to rise. Coal is still abundant. A shift back to coal is the quickest alternative, necessary but undesirable in several ways.

Where do the fossil fuels come from and why are these resources limited? Why cannot we drill deeper, explore more widely and continue to discover all that we need? To find the answers, we must go back in time, long before recorded history, long before the age of dinosaurs, to the days when the earth was much younger.

2
In the Time Before Morning

Imagine that we could have been witnesses to that very different world one sultry afternoon at the end of summer. Thunderclouds were building up in the western sky but the seemingly endless swamp was still. Stagnant pools of black water lay among the dense tangled roots of tree ferns, struggling upwards twenty feet towards the light. As the sky darkened, the gloom beneath the fern canopy became more intense; the air was heavy with a miasma of marsh gas and no birds were there to sing. Among the upturned roots of a fallen fern crept a wary romeriid, not unlike a small lizard but different by millions of years of evolution. A swirl of black water beneath a trailing horsetail hinted at a large amphibian, waiting.

Sudden wind combed through the tops of the tree ferns. A huge dragonfly, almost a foot long, dropped to escape the buffeting and swooped towards a hummock of peat as the first heavy raindrops fell. The creatures of that strange place shrank back, by instinct, as the wind began to tear at the fronds and as waves of rain swept past. One tree fern, higher than the others, strained against the force. It swayed but did not yield, though at each gust its shallow roots were loosened in the peat. Towards evening, a blast of wind greater than the rest tore the roots free and the tree-like fern fell splashing into the dark water, rolled a little and was still.

Years passed. This tree fern and others like it lay waterlogged in the black pools. They did not rot in the ordinary way; the vast quantity of debris already in the water had removed so much oxygen that bacteria and fungi could hardly survive. Gradually the softer parts were transformed to humus and the accumulated mass itself became peat, like that from which it had originally grown. Centuries passed, uncounted. Nutrient rich waters from the bare highlands continued to seep through the swamp. The recurrent cycles of growth, death, and re-birth of the dense plant life added to the layers of peat that were already there, overlying and compressing those below.

The stream of time stretched beyond comprehension to thousands and millions of years; the earth itself heaved, sank and separated under the slow but relentless tectonic forces from within. The protocontinent split apart and the Atlantic Ocean was born. New forms of life appeared and others died, no longer able to survive in a deteriorating climate or against competitors that were stronger, faster, or more prolific. The first dinosaurs were small and fleet; they and their cousins evolved into the largest land animals that have ever lived. The tree ferns of earlier times grew less densely, and the deep layers of peat in the old swamps were covered by silt and sand from the uplands, a thin sheet at first, but ever thicker. Buried, compressed by the weight of the overlying sediment, and baked by the heat rising up from deep within the earth, the peat became lignite (brown coal) and, in the course of 250 million years, the lignite became coal.

The swamps from which the coal was born had been vast in duration and in extent. For millions of years they had stretched along the tropical coastline of the great land mass that occupied much of the Northern Hemisphere. The coal that they left lay buried in thick black seams along a strip three thousand miles long. When, about

200 million years ago, the proto-continent split, the westerly portion of those ancient swamps, what we now call the Appalachians of North America, separated from the easterly basins of England, northern Europe and the Ukraine. So great was the legacy of that time that it gave its name to an age—the Carboniferous Period of the Paleozoic Era.

It is not difficult to put together an outline of the sequence involved in the formation of coal—there are examples of each step on hand today. Peat is still being formed in a few places such as the Dismal Swamp of North Carolina, though the amount is miniscule compared with the vastness of the Carboniferous swamps. Deposits of peat lie at or very close to the surface and it is lifted for use as fuel in various places like the Scottish Highlands and Ireland. It is a rather poor fuel, giving more smoke than heat which is usually a disadvantage, except, it is said, in making whiskey. The plant residues in peat are quite obvious, being only slightly modified with the passage of time.

Lignite, or brown coal is the next step in the sequence. It may be quite old, even on the vast scale of time that we are scanning, but was never much compressed by overlying sediments. Still poorly consolidated as a rule, plant residues in it are again easily apparent to the eye. Bituminous coal, which constitutes the vast bulk of the ancient deposits, has been buried deep in the earth, is black, dense, and brittle; plant structures are usually visible in bituminous coal only with the aid of a microscope, while anthracite, very hard and brittle, generally contains no recognizable plant residues.

The other great organic fuel is of course oil and the natural gas with which it is closely associated. It is perhaps rather surprising that no geologist knows with certainty how oil was formed in the first place. One problem is that oil and natural gas, being fluids, can percolate through a porous rock often for considerable distances until they reach either the surface (and evaporate) or some impermeable stratum. Oil and natural gas are found not where they were formed, but where they are trapped. Chemistry gives us some clues, but not very good ones. Oil is certainly derived from organic remains, probably marine, but whether from plants or animals is unknown. Our ignorance is as great as that. We can however make some speculations. Maybe the plants or animals (or both?) were deposited first as marine ooze, worked over many times in the digestive tracts of the animals of the sea floor before final burial, but the manner of transformation of organic matter into oil eludes us. It is, on the other hand, reasonably certain that oil and coal are *not* the liquid and solid end products of the alteration of peat; the formation of oil seems to be essentially different from that of coal. Little free oil has been found from rocks that were formed from sediments in fresh water. The vast oil deposits of the Middle East must have been produced in shallow seas that teemed with life for millions of years until the gradual rifting of the earth's crust and the redistribution of the continents brought it to an end. The layers in which the organic debris had accumulated were covered by finer, impenetrable material, perhaps silt that became shale, and again altered by pressure and heat. One other thing is certain, though. As with coal, the process of oil formation is very slow—no oil has been found in sediments that are less than 15,000 years old, even though they contain organic compounds that may be embryonic oil. The process certainly takes longer than that, perhaps hundreds of thousands of years.

The search for oil is, in the first place, the search for geological structures that can collect oil. Marine sediments, when first laid down, are usually in horizontal layers; they are porous and saturated with water. The passage of time may bring variations in climate and sea level and consequently differences in the type of deposition. Different strata accumulate one above the other. As oil is formed, being lighter than the water, it gradually separates out, percolating upwards until it meets an impermeable layer above. There it can be trapped at the highest points, with salt water saturating the rock below. Gas may be dissolved in the oil or it may also accumulate at the top.

The relentless working of the earth deforms the layers, folding them, heaving them up in places to form mountains and the oil or gas, if there is any, remains trapped at the crests of the folds. The great oil fields of Saudi Arabia and Bahrein are found in structures like these, but in many others, whose form is very similar, there is no oil at all.

A quite different type of geological structure can also trap oil. A buried large mass of light material, such as rock salt left behind by the evaporation of shallow seas, will gradually force its way upwards

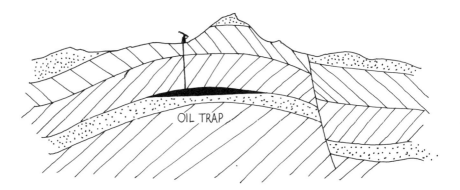

like a bubble rising to the surface of honey (but taking many millions of years to do it), and forming what is known as a salt dome. As it rises at a rate that is, within our human time scale, imperceptible, it bends the surrounding strata upwards, forming pockets that again can collect oil. In Texas and Louisiana, both on-shore and in the Gulf of Mexico, salt domes like this have been rewarding sources of oil.

The modern prospector has at his disposal instruments that can delineate the underlying geological structures without the necessity for drilling. Sound waves from an explosion travel downwards and are reflected from the various strata in the same sort of way that an acoustic depth-sounder on a boat measures the water depth by reflecting a sound pulse from the bottom. The greater the time interval between transmission of the pulse and the reception of the reflected echo, the greater the depth of the water. The oil prospector may be faced with a series of reflecting strata each of which produces a separate echo of different strength, but the basic idea is the same. He can plot out the geometry of the strata buried beneath the ground, looking for those structures in which oil *may* have collected. He does not know that any oil is there until he drills, and herein lies the hazard of wildcatting, even when supported by the best scientific information available. Maybe the driller will find oil, maybe he will find gas, but usually he will find nothing. Most wildcat wells are dry. Where the oil has come from is of little concern to him; he does want to know where it might have been trapped, but most of all, he wants to know whether any is there.

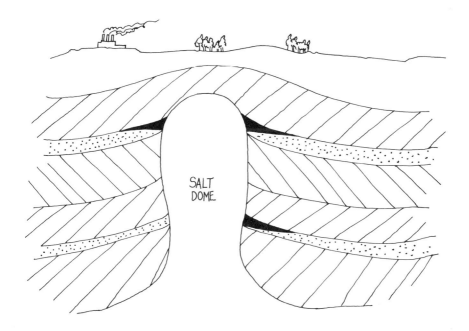

Coal, oil, and natural gas are organic fuels; the story of nuclear fuels like uranium is very different, but also fascinating. Uranium in rocks found at the present day has always been there—in fact, when the earth was formed, there was very much more. Uranium atoms decay spontaneously, ejecting sub-atomic particles which produce heat as the uranium undergoes a series of transformations, ultimately becoming lead. Over the eons this process and others like it have made the heart of the earth as hot as it is. Uranium does not burn chemically like coal or oil. In the fuel rods of a nuclear power plant, the uranium content is vastly enriched, and energetic sub-atomic particles induce the break-up or fission of neighboring uranium atoms with which they collide. The nuclear reactions can be controlled and sustained; a great deal of heat is produced which provides the energy source for the power plant.

The question of why the primordial earth had just so much uranium (and so much oxygen, hydrogen, gold, iron and each of the other elements) is a profound one; to try to answer it would involve speculation about the origin of the solar system and the universe itself. Let us simply accept the fact that, when the earth was formed, uranium was present, and although most of it has since decayed, a certain amount remains. A more easily answered question might be: How did it come to be concentrated into deposits that could be mined rather than being dispersed more or less uniformly as a rare constituent of the earth's rocks?

Even this apparently simpler question, asked not only of uranium but of any mineral deposit, contains many of the riddles that have perplexed geologists for a century or more. Most natural processes are dispersive—the wind scatters smoke from a fire; excess fertilizer on a field is washed out and carried by streams to a distant estuary. How is it that great mountains of iron ore, almost pure, could have been formed in Minnesota, Western Australia and various other places? Or gold deposited as nuggets or thin seams in a matrix of quartz? The iron, the gold and the uranium were present in the material from which the earth was formed, but the concentrations of them certainly were not.

It is only in the last few years that we have begun to understand how these aggregations have occurred. It was mentioned previously that the earth is not a static thing. Two hundred million years ago the proto-continent that was to become America, Eurasia, and Africa split apart and gradually separated, until today the Atlantic Ocean is some four thousand miles across. The idea of drifting continents is not new. It was suggested during the early years of this century by the German geophysicist Alfred Wegener, who was intrigued by the remarkable coincidence between the shapes of the coastlines on either side of the Atlantic—they seemed to fit together like a jigsaw puzzle. At that time, his suggestion had little good evidence to support it and was generally dismissed as (to quote one eminent geologist of the day) "not scientific, but (taking) the familiar course of an initial idea, a selective search through the literature for corroborative evidence, ignoring much of the facts that are opposed to the idea, and ending in a state of auto-intoxication in which the subjective idea comes to be considered as an objective fact."[1] In the face of such attacks, the notion fell into disrepute for almost fifty years. Since

[1] *Theory of Continental Drift, a Symposium* (Tulsa: American Association of Petroleum Geologists, 1928).

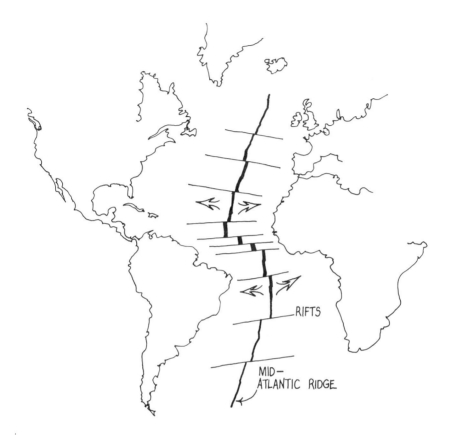

World War II, however, newer techniques for mapping the sea floor have shown that the jagged line of the mid-Atlantic Ridge, stretching from Iceland almost to Antarctica, far below the surface of the sea, is quite unlike any mountain chain on land but is, in fact, the tearing line on either side of which the sea floor is opening out. The process is still going on today, the pattern of distant earthquakes that can now be detected along the ridge and their intensity indicating clearly that the continents are still separating, gradually, to be sure, and too slow to be measured directly, but at an average rate of one or two inches a year. Though Wegener did not live to see it, his apparently crazy idea has become an inescapable conclusion that revolutionized our conception of this dynamic earth.

By the late 1960's, we realized that sea floor spreading is not unique to the Atlantic Ocean but occurs also along submerged ridges that divide the Pacific and Indian Oceans and the waters that circle Antarctica. New sea floor is constantly being created along these lines, moves slowly to one side or the other and ultimately disappears again into the bowels of the earth in the great oceanic trenches along Japan, the Philippines and South America. Not only that, even now a proto-ocean is gradually forming as the bed of the Red Sea laboriously sunders, with Arabia and Africa moving apart, imperceptibly but relentlessly.

At the bottom of that sea, in hot, deep basins, new mineral deposits are being formed—not quite before our own eyes but certainly as we observe that hostile place with deep searching instruments. In deep basins of the Red Sea lie pools of dense, hot water, heavy with dissolved minerals. It is not ordinary sea water—the composition of its dissolved solids is quite different—but ground water that, throughout the years, seems to have percolated down from the highlands of Ethiopia, dissolving minerals on the way and then finding itself

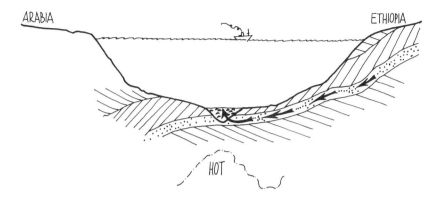

heated by the great intrusive dike beneath the sea floor that is forcing its way upwards. There is no oxygen in these stygian pools at the bottom of the sea—their water is dense with sulfides and heavy minerals and devoid of life. As the water eddies upwards, it encounters the lighter, oxygenated water above. The line of demarcation, the surface of the pool, is quite sharp and remains so as the dissolved sulfides are oxidized; the heavy metal oxides precipitate out, raining back as flocculent particles upon the sea floor beneath. The metallic oxides—proto-ores, if you will—are being formed today; with time, thousands upon thousands of years, a large deposit will have accumulated. The metals that were dispersed throughout the rocks of Ethiopia are being concentrated at the bottom of the Red Sea. With some future upheaval of the earth, in perhaps one or two million years, the deposits may become terrestrial, a prize for that distant time if anyone be here to claim it.

This is the kind of process that during ages past produced the concentrations of ores, including uranium, that we have so busily and so quickly exploited. The rate at which we have extracted them (and often re-dispersed them as trash) is so rapid compared with the gradual and patient way in which they were formed that, on the scale of the lifetime of the earth, it is taking a mere blink of time. What exists for this generation and those that follow is what we have now. During the lifetime of our civilization, there is, in effect, no more.

3
The Profligate Age

The fossil fuels slept untouched for hundreds of millions of years. In the last moments of that time span, man appeared, a strange, rapacious, thinking creature who began to use for his own ends the materials that he found about him. He discovered flints and learned to make tools of them. He discovered fire, burned wood and found that the heat reduced certain stones to metals which he could mold and shape. He built with wood, stone, and sweat. The short swords of Caesar's legions were forged with the heat of wood and charcoal.

Wood sufficed for heating, cooking and the beginnings of industry for another fifteen hundred years, but in Tudor England the first energy crisis did develop. It was a crisis in slow motion but it bit deep. By 1540, the local woodlands in eastern England were becoming depleted, and the price of firewood doubled and doubled again by 1580.[1] The fuel crisis hit the poor first and hardest; city dwellers took

---

[1] A. D. Dyer, "Wood and Coal; a Change of Fuel," *History Today*, September 1976, pp. 598–607.

to pillaging the surrounding countryside, tearing down fences and ripping out hedges. By the end of Elizabeth's reign, the crisis was general. Legislation was passed to conserve the woodlands that remained, but efforts at conservation were only palliatives.

A new source of energy was needed and it was found in coal. Until the late sixteenth century, the term "coal" had meant charcoal; the new fuel was called "sea coal." Since the Middle Ages, its use had been a local oddity, confined mainly to blacksmiths living near accessible outcrops. The increase in price of firewood and charcoal stimulated the mining of coal at Newcastle and in South Wales; the British coal production rose from perhaps 50,000 tons per year in the 1540's to at least one million tons in 1680, an expansion without parallel in preindustrial times.

The switch to the new fuel brought its problems. John Evelyn, the diarist, wrote in 1661[2] that the smoke in London caused such a stench that a traveller "sooner smells than sees the city to which he repairs." The London air "scatters and strews about those black and smutty atoms upon all things where it comes, insinuating itself into our very secret cabinets and most precious repositories." But the new fuel came just in time. In 1698, Thomas Savery built the first practical steam engine for pumping water from a well, and the designs were refined by Newcomen and Cowley in 1705. By 1763, James Watt had built engines powerful enough to drive high efficiency machines in factories and weaving mills, and their furnaces became ever more hungry for coal.

Railroads in 1800 were used for hauling coal from mines, but the motive power was provided by horses. The first reasonably successful locomotive was built by Richard Trevithick and Andrew Vivian in 1804, and the next eighty years saw the explosive growth of railroads throughout Britain, the United States, and Europe—practically to the ends of the world. The Baltimore and Ohio Railroad, built by a group of Baltimore merchants to compete with the Erie Canal, ran its first locomotive in 1829, the Stourbridge Lion, imported from England. In a celebrated race against a horse that year, the locomotive lost. Nevertheless, within a very few years railroad lines were snaking

[2] An interesting account is given by S. R. Smith, "John Evelyn and London Air," *History Today*, March 1975, pp. 185–89.

through the Cumberland Gap to the west, and the feeble, cranky engines had become giants. Coal was needed in larger and larger quantities to power the industry that was developing in the East and to drive the locomotives and steamships carrying goods and settlers to the opening West. Already, the burgeoning development of our country was becoming dependent on its base of abundant energy.

Other forms of energy appeared. In England, by 1831, the self-educated Michael Faraday had discovered electromagnetic induction which made possible the electric generator and electric motor. The first U.S. central generating station and power lines were erected by William Stanley in 1886 at Great Barrington, Massachusetts. The generators were turned by steam, and the furnaces burned coal. The energy web spread. Seepages of rock oil had long been used in linaments and internal medicines, but the oil well drilled in 1859 at Titusville, Pennsylvania, led to the world's first oil boom. Those great deposits that had lain dormant for millions of years were soon to be pumped out; their moment had come. With the advent of the automobile at the beginning of the new century, gasoline became a major product, and the energy spree accelerated.

Stimulated by the Civil War and the two world wars of this century, American industry became a colossus of unparalleled power and with an unparalleled appetite for energy. The extraction of metals from ores as well as refining and fabrication depended on cheap and abundant energy supplies; manufactured articles were no longer hand-crafted but mass produced. As each unit of time became more productive, each bit of time saved seemed to be worth more. There was a steady decrease in the value of *things* compared with the cost of a man's time; all but the poorest families could afford a sewing machine and later a refrigerator, an automobile, and a television set. When appliances broke down, it was often cheaper to junk them and buy new ones than it was to have someone search through for the problem and make a sometimes trivial repair. Junking them seemed cheaper and better for the economy: it kept the industrial plants in full production and the steel mills rolling. As long as cheap and plentiful energy was available to provide a base for it all, we could build automobiles that would fail within five years and refrigerators that would fail in ten because we needed to sell all the new ones that industry could produce. Our material standard of living rose in bounds.

People became not persons but *consumers*. Agencies were set up to protect, not people, but consumers. It was patriotic to consume, slightly un-American to fix things yourself. We became so good at consuming that our cities began to run out of space for the trash, and it invaded the surrounding countryside. Towns became ringed with piles of junked automobiles. Sometimes, but not often, it was worth re-cycling the metal. While ore was plentiful and energy cheap, it was less expensive to produce steel afresh and leave the junk to rust. The profligate age had come.

There was a certain ethic in all of this, an apology for materialism that could be made. It was fallacious but perhaps superficially plausible. In one sense, the worth of man was enhanced relative to that of material things; a part of a person's lifetime became of greater value than a refrigerator. We could be freed of much of the drudgery of manual labor, our activity had the potential to be more creative, of more service to our fellow beings in a human, less mechanical way. Perhaps, to an extent, this happened, but the potential was never realized. Leisure-time pursuits of various kinds came to flourish, but so did passivity. Television became a great merchandising industry as it seized upon a massive and supine audience.

Still, the machines whirred, the automobiles streamed along the new freeways, and the energy was there to support it all. We were beginning to import an increasing amount of energy, mainly oil, as our appetite outstripped our resources, but it hardly mattered while imported oil was cheap. Our material standard of living was rising steadily and there seemed to be no end in sight. We were a people truly blessed.

Then came the first great public shock—the oil embargo of 1973.

4
Growth and Decay: The Inexorable Cycles

It is, of course, no surprise that non-renewable resources do eventually run out. In Phoenician times, the tin mines of Cornwall supplied much of the ancient world, but there is little tin in Cornwall now. The middle of the last century saw the great gold bonanza in the West, but the gold fields today attract only tourists, not miners. The same is certainly true for oil in the United States—only a finite quantity of it was there in the first place and our exploitation will certainly, some day, come to an end. But as long as "some day" seems sufficiently remote, who would care to worry?

In the careless days, to be concerned about depletion of national or world resources was to risk being called a prophet of doom not to be believed. Fortunately, however, a few thoughtful people were deeply concerned. As early as 1956, the American geophysicist M. King Hubbert startled the petroleum industry by showing with brutal clarity that the depletion of oil in the United States was following a clearly defined pattern, and that exhaustion of the resources was not distant, but predictably near.

Outside geological circles, the name King Hubbert is not widely known even today. The facts of his life are not remarkable. Born in 1903 in central Texas, he attended the University of Chicago, receiving his Bachelor's and Master's degrees by the age of twenty-five. He held a junior post at Columbia University for ten years, during which time he completed his Ph.D. degree from Chicago. Much of his subsequent career was with the Shell Oil Company and the Shell Development Company while holding a visiting professorial appointment at Stanford University. He retired from Shell in 1963 but has continued to work as a research geophysicist with the U.S. Geological Survey. What is remarkable about the man is the independence of mind and the clarity of thought that brought him ultimate honor among his peers, culminating in 1977 in the receipt of a Rockefeller Public Service Award. The same qualities made him a maverick.

He had been concerned with the patterns of discovery, exploitation, and depletion of non-renewable resources of various kinds, but particularly petroleum. It was not dramatic or exciting work; it involved sorting out and collating production figures, poring over tables of numbers, charting, analyzing. Piece by piece, however, a pattern began to emerge. This pattern transcended the nature and the extent of the resource. Even technological innovations that might be thought to revolutionize an industry influenced the pattern only slightly. He found that the discovery, exploitation, and depletion of these non-renewable resources appealed to follow predictable and inexorable cycles, marked by a beginning, a climax, and an inevitable end. Once the pattern was seen, its simplicity was breathtaking, its implications startling and clear.

Let us imagine, for the sake of a simple illustration, that many years ago a treasure ship was wrecked on a beach. The ship broke up and the treasure was scattered and buried up and down the beach. In some places there are quite big aggregations of gold coins while in other places there are just one or two coins, widely scattered. This is our non-renewable resource—there is only a finite amount of it, the amount that was on the ship in the first place. One day, you are walking along the beach, and you happen to pick up a gold coin. Greatly excited, you search around carefully, perhaps dig a little and find another one. This is the initial discovery phase.

Now you become more serious about it. You buy a metal detector and start to search more diligently. As you work, you locate other promising places; in fact, you begin to discover spots where gold coins might be, almost faster than you can dig them out. So you hire someone else to do the digging for you in places where the indications are especially promising. As you become smarter at the task of finding the gold, you succeed more often; you develop a feel about the way the whole treasure has been scattered along the beach and your rate of discovery increases. Meanwhile, the rate of recovery by digging lags behind—*you are discovering it faster than it can be dug out*. Your reserves—the amount discovered but not yet extracted—are increasing. But all the time, you are discovering more and more of the total resource, the finite amount that was there in the first place, and then there comes a time when *your rate of new discoveries no longer climbs, but peaks out*. There is still a lot more to be found, but it is buried deeper, or more widely scattered, and no matter how hard you work with your metal detector, new discoveries start to come more slowly simply because there is less and less of the treasure left for you to find. Meanwhile, there is still the backlog of treasure already discovered but not yet dug out, so that production keeps climbing even though your rate of new discovery is continually sagging. *The rate of production lags in time behind the rate of discovery*. Your reserves begin to decline. As time goes on, you approach the end of your discoveries. You have found almost all there is; you can work night and day with the metal detector, but you are not going to find much more. By this time, the rate of production is declining also since most of the easily accessible treasure has already been dug out, and the remainder is more and more difficult to extract. You may be obliged to rent more elaborate machines to do the digging, and the costs rise. In the end, the rate of discovery drops off to the point where it is no longer worth the search. There is only the backlog to be dug out, and when that is finished, you are through.

The history of your venture can be illustrated by simple and revealing charts that plot either the rate of discovery and the rate of extraction day by day as the treasure hunt progresses, or else the total amount discovered and extracted to date. The rate of discovery cycle starts with the initial phase when the first few coins are found, rises to a peak when the rate of new discovery is as large as it is ever going to be, then drops as new finds become more and more difficult, and finally tails off to almost nothing. The rate of production cycle lags behind, but also rises to a peak as the backlog in discovery is being exploited. Then it, too, begins to drop off as the treasure becomes depleted, and the remainder is more difficult to dig out. The total amount of gold discovered in the search is the rate at which you find it, added up day by day throughout the cycle; in the graph, it is represented by the area under the discovery curve. If you can extract

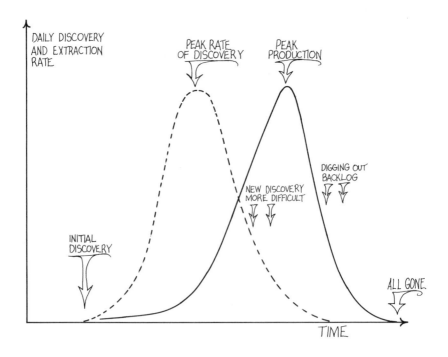

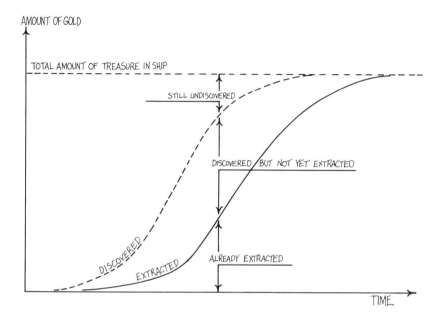

all that you discover, this must also be equal to the total area under the production curve.

The same cycles can be shown in a different way by charting the total amount of gold discovered and extracted to date as time goes by. The total amount that was there in the first place does not change with time; it is represented by the horizontal line at the top. After the initial discovery, the fraction of the total that you have found rises with time as your exploration proceeds until, ultimately, you have found nearly all of it. (You may miss some—there comes the time when it is no longer worthwhile to search any more.) The production line lags behind; the amount that you have already dug out keeps on increasing as you work at it until ultimately it, too, approaches the total amount that was ever there. But the gold must be discovered before it can be extracted. The *difference* between the two curves at any particular time represents the gold that you have discovered but not yet dug out—it is your proven reserve. It is small in the beginning, then it rises, remains approximately constant for a time while you are discovering and producing at about the same rate, and then it declines as the rate of production exceeds your rate of new discoveries.

These are the cycles that Hubbert perceived, the separate cycles of discovery and production that are involved in the exploitation of any finite natural resource. Two points about them are extremely important. First, the discovery cycle leads the extraction cycle—the rate at which the resource is extracted lags in time behind the rate at which it is discovered. The second most important factor is that the discovery cycle in time is almost symmetrical—the time that it takes to rise to a peak is about the same as the time taken to fall from the peak to the point where essentially no new discoveries are being made. When the rate of discovery is a maximum, this indicates that you have already found about half the total amount of resource present, that there is still only half left to be found. As long as the rate of new discovery is still climbing, we do not know how big the bonanza is, but as soon as the rate of new discovery peaks out, *that is a clear warning that we are about halfway to the end.* As time goes by, the new discovery cycle begins to decline, even though the production cycle is still rising. Proven reserves begin to decrease as production approaches its peak and outstrips the rate of new discovery. The more accessible, richer deposits of the resource become worked out, and we are forced to turn to the deeper or more difficult ones; production becomes more expensive, and the rate declines steadily.

By this time, it is obvious that the industry is decaying, and the end is in sight. But, as Dr. Hubbert realized, the signals had been clear much earlier. Once the rate at which new discoveries were being made had reached a peak, only about one half of the resource remained to be found. Even though production was still rising at this point, the peak would inevitably come some years later, and after that, the decline would be inexorable.

These Hubbert cycles appear to apply universally to large-scale, well-distributed non-renewable resources, whether sapphires in Sri Lanka, oil in the United States, or oil in the whole world. Can they be stretched out by new discoveries? No, the new discoveries are already taken into account; they are provided for, even though not yet made. If, towards the end of the cycle, increases in price stimulate the rate of search, new discoveries will be made, but in general they will be smaller, deeper, more difficult and more expensive to extract.[1]

Hubbert's cycles are, at the heart, statistical in nature, and so are applicable only to resources with many individual elements, many deposits, many mines, or oil wells. With large numbers, averages become more accurate and predictable; with small numbers, they may be less significant. An insurance actuary can tell with precision the proportion of those people listed in a city telephone directory who will be alive one year hence; he can offer no guess on any individual. In the same way, the Hubbert cycles are accurate and inexorable in predicting the exploitation of resources that consist of many elements, among which the individual fortunes of a particular oil well or oil field are but a part of the whole.

Sometimes the cycles can be distorted a little. They are not always so clear in the case of small, localized deposits in which the extent of the resource may be established quickly and the production rate can be controlled readily, increasing or decreasing until the depletion point is approached. Sometimes an improvement in technology may make it possible to recover material from a deposit that was previously too low in grade to be worthwhile. Sometimes an increase in price can stimulate a decayed mining industry into fitful further production, but the end result is inevitable. The overriding fact is that a finite amount of the resource was present in the first place, and when it is extracted, there is no more.

Unfortunately, the example of the treasure is hypothetical. It may make eminent sense, but does it, in fact, work that way?

In 1956, King Hubbert constructed a graph of the new discoveries of crude oil in the United States by year since 1900, a graph that was to shake the petroleum industry to its foundations. It showed the rate

---

[1] Proponents of deregulation of natural gas prices emphasize (correctly) the fact that new discoveries will be made, but cheerfully ignore the second (equally correct) part of this statement.

of new discoveries of oil in the United States, year by year from 1900 to 1956, and looked like this:

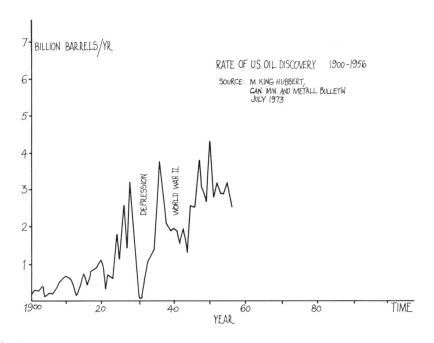

At first sight, the graph is quite irregular—new discoveries in a particular field tend to come quickly once the field is located, after which there may be a hiatus. There are clear dips during the great depression, when nobody could pay for oil exploration, and during World War II, when there were other more immediately pressing matters. What caught Hubbert's eye, however, was the fact that during the ten years prior to 1965, the rate of new discovery, though jumping about, no longer showed the upward trend that it had followed till about 1945. The trend had levelled out. The zigs and zags on the graph made it somewhat confusing, but if one takes the same data and averages it over three year intervals, it becomes much smoother. Now the trend is much more obvious, and Hubbert realized the immensity of the conclusion conveyed by this simple, irregular line. *The rate of discovery of crude oil in the United States*

Growth and Decay / 35

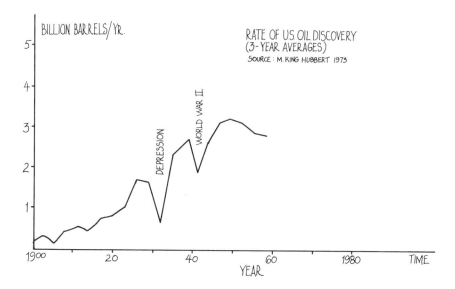

was levelling out—we were at or just beyond the peak of the discovery curve. Despite the fact that the production curve was still climbing steadily, the discovery curve was peaking, even though intensive search for oil continued. This meant two things. First, we had already discovered about half the crude oil in the United States that was there in the first place. Secondly, and most ominously, the production curve, lagging behind the discovery curve, would itself inevitably peak in the near future. Beyond this, it also would decline, no matter what the industry might do, no matter what incentives government might provide.

The crucial question was—how far in the future? How far in time does the production curve lag behind the discovery curve? This can be found by plotting (overleaf) the two curves on the same graph. If we draw smooth dotted lines through the two graphs, we see that the rate of production curve is indeed the same shape as the rate of discovery curve, but displaced to the right (towards the future) by about ten or fifteen years. The rate of discovery curve was peaking in 1956—the rate of production of crude oil in the United States would peak between 1966 and 1971, thereafter to decline. It did not matter that there was a great deal of oil still to be found—the total of the

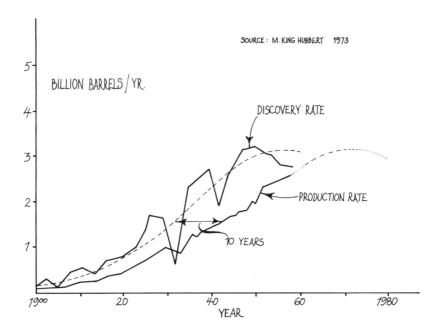

discoveries made until 1956 was about ninety billion barrels and there was about the same amount yet undiscovered. Once the rate of discovery curve peaked and began to decline, the warning was there for all who cared to read it.

The fact that the rate of discovery curve was peaking when the total amoung of crude oil found up to that time was about ninety billion barrels, pointed to a total quantity of crude that had ever been deposited in the United States of about twice this size, or about one hundred eighty billion barrels. That's it; that's all there is to be found, no matter how hard we search.

Hubbert's reluctant conclusion could be checked in a quite different way, and of course he did check it. The traditional way of estimating the total amount of petroleum present in a given area is extremely crude, but better than nothing. Suppose one locates by surface reconnaissance and mapping an area, yet undrilled, that is geologically similar to another area that has already produced oil and gas. On the basis of the similarity, strictly by the analogy between the two areas, one assumes that the virgin territory will eventually produce comparable quantities of oil and gas per unit area (or per unit volume of the sediments) as has been found in the productive area. This analogy is extremely risky—the virgin territory may have no oil at all—and has led to vast errors in the past. For example, in 1953 the U.S. Geological Survey, on the basis of the geological analogy between the on-shore and off-shore regions of the Gulf coast, estimated that four billion barrels could be recovered off the Louisiana coast. After some twenty years of intensive drilling, the discoveries of crude oil in this area amounted to some five billion barrels[2]—pretty good. On the other hand, the same method predicted that nine billion barrels were to be found off the coast of Texas. The same period of intensive drilling in this region uncovered essentially none. It is precisely this method of geological analogy that underlies the widely divergent estimates of the amount of oil that may be recovered from the Atlantic continental shelf—the experience of Louisiana and Texas illustrates clearly the huge uncertainty of such "estimates," no matter how authoritative the source from which they come.

Nevertheless, you win some, you lose some. In an area as large as the whole United States, one would hope that the errors might approximately cancel out and by adding up the amounts estimated by analogy one might hope to obtain a reasonable estimate for the total amount of oil that could be recovered from the whole country (including what has already been extracted). In 1956, the estimates made in this way ranged from one hundred and fifty to two hundred billion barrels,[3] quite consistent with Hubbert's estimate of a hundred and eighty billion, found by an entirely different method.

Everything fitted together. In March of 1956, King Hubbert presented his conclusions in an invited address before an audience of petroleum engineers at a meeting of the Southwest Section of the American Petroleum Institute at San Antonio, Texas. Consider the context: The petroleum industry had been flourishing and expanding for nearly one hundred years, demand and production were still rising, the oil companies were (and still are) one of the basic supporting blocks of our economy and among the blue chip investments of Wall

---

[2] M. King Hubbert, *Canadian Mining and Metallurgical Bulletin*, July 1973.
[3] Not counting Alaska, which may add another twenty or twenty-five, at most.

Street. Hubbert, in his undramatic way but with clear logic, was telling his audience that the peak of U.S. crude oil production would occur within ten or fifteen years and thereafter it would decline. Hubbert was not a crank, he was a respected, careful, and thorough scientist, presenting his reluctant conclusions.

His predictions were both surprising and deeply disturbing to the whole United States petroleum industry. *Somehow, he had to be wrong.* But how?

The only possibility was that the amount of oil ultimately recoverable was seriously underestimated and the sagging off, evident in Hubbert's rate of discovery curve, was a temporary lull on the way to even higher peaks. What happened next is, in retrospect, hard to believe. A flurry of new "studies" was made to estimate the total recoverable oil in the United States, with essentially no more basic information than was available before. Manipulating the same old data, the published estimates escalated to successively higher values, 204, 250, 372 and eventually an astounding 590 billion barrels.[4] Apparently, even the experts were prepared to believe what they wanted to believe. If the last figure were to be correct, there would be adequate reserves until well into the twenty-first century—hardly yet cause for concern. Hubbert stood his ground, essentially alone.

Who was right? Was he, or were the experts of much of the Geological Survey and from many of the oil companies? With twenty years of hindsight, we can now see; let us look at what has happened since then. First, we will bring the rate of discovery curve up to date, as shown on the next page.

A big peak in 1970 marks the Alaskan oil finds, but in spite of this, the trend is now inevitably down. The levelling off in the rate of new discoveries in the 1950's has become a precipitous decline and there is no question that the peak has passed. Next, let us bring the production curve to 1977. This is shown at the top of page 40.

It is chillingly evident that Hubbert's prediction was right—production did peak in 1970, just about when he said it would, and it has declined year by year since. The flow from Alaska will not reverse the decline; we can anticipate arresting it for a year or two, but after that,

[4] A. D. Zapp, *Future Petroleum Producing Capacity of the United States*, U.S. Geological Survey Bulletin no. 114–2 (1962).

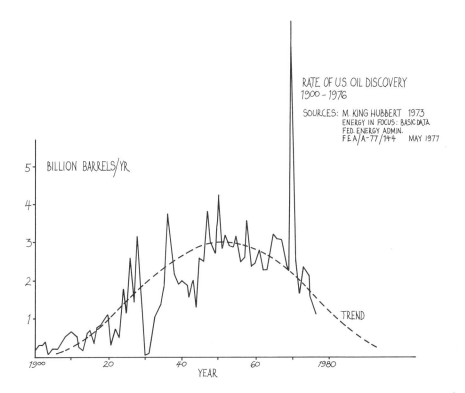

it will resume its dismal course. The fact is that between the time the Alaskan fields were discovered and put into production, the other oil reserves declined by more than the total amount that had been found there.[5] We are no further ahead—in fact we're behind, since we are using oil at a much faster rate than in 1970.

Put the two curves together and there can be no doubt about it—the rate of production curve is lagging behind the rate of discovery curve by ten years or so; we are now extracting oil twice as fast as we are discovering it, and the future course of production is disconcertingly apparent. Future improvements in recovery techniques may stretch things out a bit, but that's all.

[5] Energy in Focus: Basic Data. Federal Energy Administration A-77/144, May 1977.

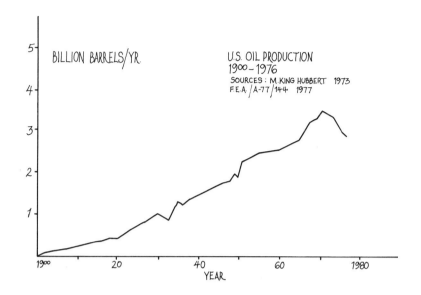

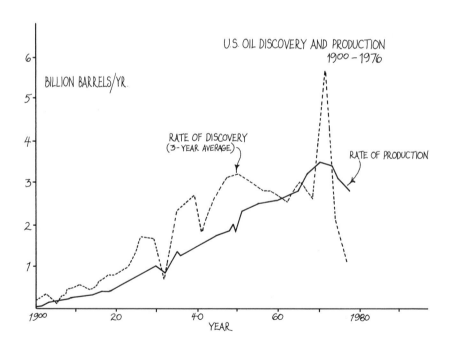

The United States is running out of oil, running out very quickly, and when it is gone, there is no more. Depletion is forever.

Hubbert was right.

How was it that many other experts were so spectacularly wrong? The extent of the country's oil reserves was not a matter of purely academic interest to be debated at leisure. It was a question of cardinal concern, not just to the industry, but to every person living in this country. The testimony of some experts, presumably given in complete seriousness and honesty, was at the same time so incredibly wrong that it led us, government and people alike, into a paradise of fools.

Nobody likes to hear bad news, and when the estimates for total ultimate production were successively revised upwards, the figures were accepted with relief. The ultimate was reached when the United States Geological Survey, no less, in response to a Presidential Directive of March 4, 1961, presented to the National Academy of Sciences the estimate that the total recoverable oil in the United States, including that already produced, was five hundred and ninety billion barrels—enough, even without imports, to supply the United States for eighty more years at the present rate of consumption. This was by far the highest estimate that had hitherto been made, and was at considerable variance with any data from the petroleum industry. The Academy had the good sense to reject it.

Nevertheless, it is interesting to see how this figure was reached, for it illustrates the way that, when statements are repeated, initial uncertainties and qualifications are often lost. Fragmentary data and tentative assumptions can lead to a suggestion that, by repetition, becomes a conclusion which is ultimately accepted as an undisputed matter of record, engraved in marble. Perhaps there is a hidden element of wishful thinking involved, again of believing what one wants to believe, that has always been a human failing. The estimate of five hundred and ninety billion barrels, offered by the Geological Survey, was based on a hypothesis introduced by Dr. A. D. Zapp, who was working at the Survey at that time but who has since died. It was a hypothesis that was even at the time breathtakingly naive and is now demonstrably wrong. Yet it was accepted at face value.

Zapp considered that the exploration for petroleum in the United States would be complete when all the potential petroleum bearing sedimentary basins had been drilled with an average density of one well per two square miles.[6] Fair enough, maybe. He calculated that this would require five billion feet of drilling. By 1961, the date of his estimate, there had been a total of 1.1 billion feet of drilling, which had led to the discovery of about one hundred and thirty billion barrels of oil. The first billion feet of drilling had uncovered one hundred and nineteen billion barrels of oil. Now here is his assumption—that future drilling will be as productive as that in the past. Consequently, he argued, five billion feet of drilling should lead to the discovery of five times one hundred and nineteen billion barrels of oil, or five hundred and ninety billion. In retrospect, it is hard to believe that this number, whose accuracy is of such cardinal importance to the country, should have been obtained in such a desultory way.

Nevertheless, it was accepted, at least enough for the Survey to present the estimate to the National Academy of Sciences. The fatal flaw in Zapp's argument is the assumption that future drilling would be as rewarding in terms of oil recovered per foot drilled, as it had been in the past. Even on the face of it, this assumption is most unlikely to be true. Oil men, one must assume, do know their trade and they do not drill at random. Drilling is expensive. They try first where there is the best chance of finding oil, and lots of it, only subsequently probing less likely, more difficult or less productive areas. The rate of discovery per foot drilled will then necessarily and inevitably decrease over time (just as with our treasure of gold on the beach) as successful exploration becomes more and more difficult. The Zapp hypothesis could only be true if wells were drilled completely haphazardly, a practice that not even the most foolhardy oil speculator would countenance.

[6] A. D. Zapp, *Future Petroleum Producing Capacity of the United States*, U.S. Geological Survey Bulletin 114–2 (1962).

The hypothesis is also demonstrably untrue. In 1969, King Hubbert (again) examined the actual records of drilling and showed that the rate of oil discovery per foot drilled, far from being constant, had already declined precipitously from the beginnings of the oil industry in this country. The first half billion feet of drilling in the United States yielded discoveries of ninety-five billion barrels of oil; the next, twenty-four billion; and the next, only seventeen billion, a dramatic indication of the diminishing return as the search for new oil continues, is intensified, and becomes increasingly expensive.

The Zepp estimate must be rejected out of hand—the total recoverable is indeed much closer to the figure that Hubbert predicted than it is to those "estimates" which followed.

Despite his success in predicting accurately the climax and subsequent decline of U.S. oil and gas production, Hubbert still has his critics. Some attack his technique by gross over-statement of it into a form that is manifestly untrue. For example, in a recent book, William Asher claimed that "the validity of (Hubbert's) method depends on the existence of a deterministic, knowable growth pattern.[7] If the approach is valid, ultimate production is precisely inferable at any point in time." This is obvious nonsense. The point that Asher ignores is one made earlier in this chapter: the Hubbert cycles are not deterministic but statistical in nature. Day-to-day and year-to-year fluctuations in oil discovery and oil production of course occur, but when many individual elements contribute to the whole, the Hubbert cycles become uncomfortably accurate in showing the inevitable and inexorable trend.

During the past few years, the U.S. Geological Survey finally discarded its euphoric predictions of the Fifties and Sixties. As late as 1974, it was estimating that there was a 95 percent chance that the ultimate recoverable crude oil in the United States, including Alaska and the continental shelves and also including what has already been produced, would be at least 300 billion barrels. However, in the following year, the estimates were revised sharply downwards. In a release of May 7, 1975, it was asserted that less than 220 billion barrels could ultimately be recovered with a 95 percent probability,

[7] William Asher, *Forecasting—an Appraisal for Policy Makers and Planners*, (Baltimore: The Johns Hopkins University Press, 1978).

and that there was only one chance in twenty that the figure would be as large as 290 billion. Secondary recovery, the application of advanced fluid injection technology to extract more oil from depleted wells, would add only about a further four billion. In terms of our overall energy supply, there is no new bonanza here.

Even the oil industry has become much more realistic about its future in the United States, though a certain amount of wishful thinking still remains. The Exxon Company offers its projections for the U.S. oil supply:[8]

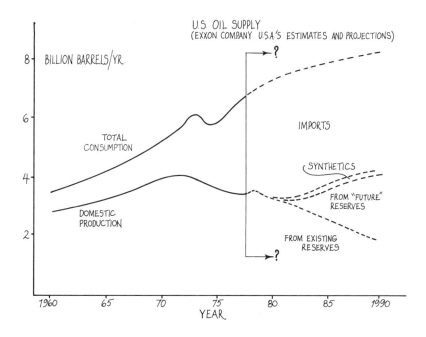

While warning that "domestic oil production peaked in the early 1970's and is declining," the curve of anticipated domestic production in the sketch above shows a tentative increase towards 1990. It is pointed out that "almost half of 1990 domestic oil production must come from reserves yet to be discovered. Most large new discoveries must come from 'frontier' areas of Alaska and the outer continental shelf." A corresponding report of January 1977 adds, more with

[8] *Energy Outlook 1978–1990* (Exxon Company, U.S.A., May 1978), p. 17.

hope than with confidence, "one hundred and twenty billion barrels may come from future discoveries and growth in existing fields." In the 1978 report, this sentence does not appear; presumably the hope has evaporated.

Remember, however, some of the awkward realities. The proven reserves in Alaska have already been used up, in essence, by declines in reserves elsewhere. The Atlantic continental shelf has maybe six billion barrels—one year's consumption at the present rate. Finally, turn back a few pages to look at the graph of rate of new discoveries: what really are the prospects of new discoveries in the 80's that can return us to the halcyon days of the 50's?

Nevertheless, it is apparent that the oil companies are becoming increasingly and uncomfortably aware of the limits of possible oil recovery (Hubbert, after all, was with Shell), and their belated efforts to promote conservation must be commended. One would only wish that equal efforts were being undertaken to provide suitable alternatives.

For natural gas, the prospects are alarmingly similar, although lobbyists for the industry would have us believe otherwise. Natural gas was originally an unwanted bi-product in the extraction of oil. It had limited use as an industrial fuel in the mid-western oil fields, but the difficulties in storing and shipping led to most of the gas being simply burned off at the well. It was conspicuous waste, but it was the cheapest thing to do. Not until the 1940's, when a pipeline network was developed to connect the southwestern and western fields to the population centers of the midwest and eastern coasts, did the use of natural gas become widespread. It is now a premium fuel, clean and relatively cheap; low gas prices, regulated by the government, encouraged conversion from coal and oil not only for domestic heating but also for electric power generation and industrial use.

King Hubbert was also concerned about natural gas. In 1962[9] he

---

[9] M. King Hubbert, *Energy Resources*, National Academy of Sciences—National Research Council Publication 1000–D (1962).

estimated that the peak production rate of natural gas would occur about 1977; in fact, it came earlier, in 1972. The prophet of doom had, in fact, been over-optimistic. He made his estimate when the rate of discovery curve appeared to be peaking out even though the production curve was still rising; let us see how the curves now look with the added information of the past fifteen years.

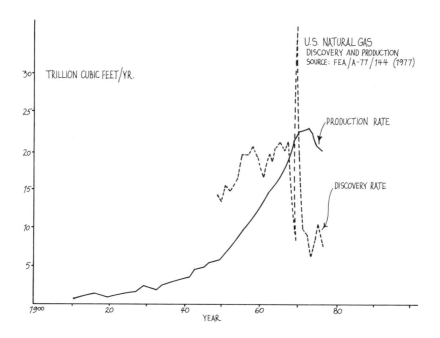

It is obvious, even at a glance, where we are today on the cycles. The rate of discovery, after a big peak representing the large Alaskan finds of 1970, has subsided to less than half its level attained in the 1960's. The production curve, lagging behind it in time, has clearly also passed its peak and is now declining.

Lobbyists for the industry have complained vociferously, in arguing for deregulation of wellhead prices for gas, that exploration for gas has been discouraged by the artificially low prices. The pricing of gas is a complex political issue, and the debate about it has been vigorous. Steven Rattner, in the *New York Times* of June 30, 1977, commented wryly from Washington: "Though bitter political battles are not unusual in this capital, the intensity of the fight over natural gas pricing has startled even veteran observers." Of course: a great deal of money is involved. The price of natural gas, like that of Middle Eastern oil, has little to do with the cost of discovery and extraction; it depends more directly on what the consumer can be made to pay. Rattner quotes the Natural Gas Supply Committee, the principal lobbying arm of the producers: "Only a bill which includes the clear and certain promise of deregulation of wellhead prices offers any possibility of bringing supply and demand into meaningful balance." The argument is that if prices are deregulated, their subsequent increase will encourage conservation while more gas wells will be drilled, and this will restore the old rate of discovery and eventual production. Is this true, or is it just whistling to the wind? Let us look at the record of the actual number of gas wells drilled since 1960:

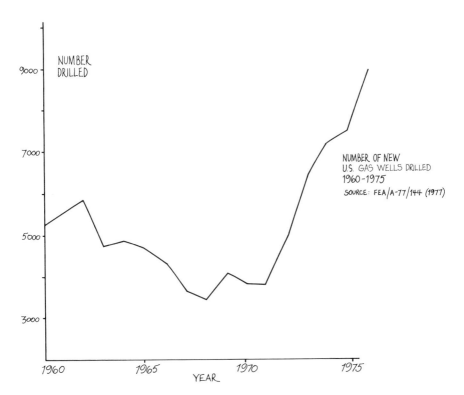

NUMBER OF NEW U.S. GAS WELLS DRILLED 1960-1975
SOURCE: FEA/A-77/144 (1977)

Note the interesting fact that the number of wells drilled in 1976 has more than doubled since the 1960's, yet turn back a page or two to the rate of discovery curve for gas to see what has actually been found in this increasingly intensive search.

The Exxon Company is more realistic in assessing the future of natural gas. In its booklet, "Energy Outlook, 1977-1990," the Company states: "U.S. natural gas production peaked in 1972 and has been declining ever since. Production is not expected to recover to 1972 levels, even with production from new offshore leases and from the Alaska north slope."

Rhetoric and political lobbying may obscure the issue, but cannot overcome the inexorable logic of the Hubbert cycles. The fundamental point the reader will now find quite obvious is that natural gas is a finite resource, non-renewable, and an increase in price is not going to expand the resource one iota. It can only make producers more enthusiastic about extracting the remainder more quickly. Conventional economic wisdom—that by raising prices, you stimulate production—may work for renewable resources like timber and corn, but for non-renewable resources it can only accelerate depletion.

There is, nevertheless, a great deal of money still to be made in the discovery and exploitation of oil and natural gas. Make no mistake about that—there are still fortunes to be made (and lost). For example, James P. Sterba, writing in the *New York Times* of November 6, 1977, describes the success of Houston Oil and Minerals, whose stock "soared on the American Stock Exchange from $2.10 a share in late 1972 to $66.00 last year, adjusting for four splits along the way. Net earnings leapt from $790,000 in 1971 to $38.4 million (in 1976).

"Why? Mostly because in the last five years, Joe Walter (President and Chairman of the company) has found three-quarters of a trillion cubic feet of natural gas and thirty to forty million barrels of oil in virtually one Texas county—a county that had been explored, drilled and otherwise picked over by oilmen since the 1930's, using outmoded and less intensive techniques. Which means, he says angrily, that those dunderheads in Washington, including the current occupant of the White House, are full of baloney when they say there isn't enough oil and gas left to justify the price incentives needed to go look for it."

Clearly, the company has been a great financial success, but is this aggressive confidence justified? The amounts of oil and gas that have made this company prosper represent big money but are, in fact, insignificant compared with our overall energy appetite. Let us put it in perspective. The national consumption of natural gas is about twenty trillion cubic feet per year—the three-quarters of a trillion discovered by this company over five years, represents a mere two weeks' supply and the thirty or forty million barrels of oil would serve our needs for less than two days.

Nuclear plants, generating electricity, depend on uranium just as conventional fossil fuel plants depend on oil or gas. The story of uranium has several different twists, but by now the reader will need little prescience to guess that the ultimate outcome can be no different. The rate of discovery of uranium oxide in the United States has had two distinct cycles (overleaf) simply because the demand arose, disappeared, and then returned again. During the 1950's, the pace of exploration built up rapidly and then declined as military requirements reached saturation and incentives were removed. The rate of discovery dropped. With the advent of nuclear generating plants in the mid-1960's, the increased demand stimulated further exploration and production increased once more.

In spite of the distortions produced by fluctuations in demand, Hubbert's logic still seems to apply with awful inevitability. The production cycle lagged behind the discovery cycle, and as the U.S. deposits of relatively high-grade sandstone ore were more thoroughly explored, new discoveries became more difficult to come by. In 1956, the discovery rate in terms of pounds of uranium oxide per foot drilled was 18.6; in 1968, when exploration was much more intense,

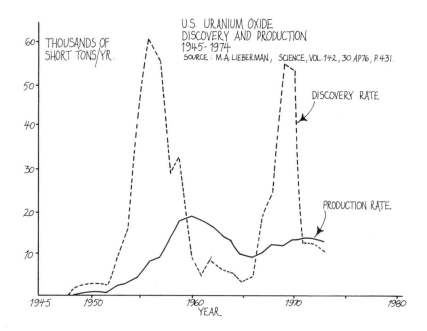

it had dropped to 6.9 and by 1973 it was only 2.4.[10] Proved reserves, largely in Colorado, Wyoming, and Texas, are still substantial, amounting to about as much as has been extracted since 1950 and, furthermore, the resource could be stretched by re-processing if it were available in time, but that raises many more serious problems to which we will return later. Nevertheless, the writing is on the wall. In 1976, M. A. Lieberman, using the same techniques that Hubbert had pioneered, estimated the ultimate recoverable resource of uranium oxide in the United States at present costs to be about 630,000 short tons. By 1975, we had already extracted 270,000.

Uranium is not a common metal—even the 'high grade' sandstone ores now being mined contain as little as one part per thousand of uranium oxide. Lower grade alternatives, western lignite and shale, contain less than one part of uranium oxide in ten thousand; the costs of extraction are correspondingly higher. As Lieberman writes, "If any significant growth in U.S. nuclear power is to be sustained over

[10] M. A. Lieberman, "United States Uranium Resources—An Analysis of Historical Data," *Science* 142, 30 April 1976, p. 431.

Growth and Decay / 51

the next few decades, then either a marked increase in the rate of discovery of new, high-grade reserves must occur, contrary to past trends in exploration statistics, or the development of low-grade ores ... must be undertaken immediately and vigorously. The alternative is dependence on foreign sources of supply which may be uncertain both politically and as to the quantities actually available. If the uranium supply does not materialize, then nuclear electric power will saturate at the levels far below those now projected for the next few decades."

Is nuclear power our long term answer?

OLD KING COAL

Now for the good news. The legacy of those carboniferous swamps was so great that we have hardly begun in the Hubbert cycle for coal. Even if we begin to extract coal at five times the present rate, there is enough in the United States proven reserves to last about three hundred years.

5
Megapower and Micropower

There is little enough comfort to be found in the stark recitation of the previous chapter. The facts are from hard historical records; they are not doomsday "estimates" or "projections." The problems face us squarely. If they are to be overcome (and with resolve, they can be), their true extent must be appreciated by us all. Churchill did not prepare the British people for the Battle of Britain by telling them it would be easy. The energy battle can be won, but it will not be easy either.

Why have we not been told long since? A cynic might reply: Who is to do the telling? The politicians? More often than not it seems that their first concern is to have themselves re-elected—these problems cannot be solved before the next election. The energy companies? Perhaps, but their purpose is to distribute and sell energy, not to warn of massive impending shortages. But the cynic may be less than fair; some voices have spoken out. Some utility companies and oil companies, particularly Exxon, have made a few belated, often half-hearted attempts to encourage conservation. Attention by the media has been sporadic and sometimes clouded by misinformation. True, some environmental organizations *have* spoken out, but their constituencies are limited. In one unlikely source, *Fishing Facts Magazine* of November 1976, an editorial article by George Pazik gave an unusually blunt and generally accurate appraisal of our energy prospects. But this magazine is not among the regular reading of most American people. In spite of attempts such as these, the magnitude of the problem is not widely known. The plain fact is that, in spite of occasional temporary let-ups, the energy crunch is much worse than is generally believed, and very much worse than most industry lobbyists would have us believe.

King Hubbert, of course, spoke up whenever he had the chance, but he spoke quietly and mainly to technical audiences and through the scientific journals. As long ago as 1957, Harrison Brown, a highly respected professor of geology at the California Institute of Technology, writing with James Bonner and John Weir in a book entitled *The Next Hundred Years*,[1] warned of the end of the petroleum era. "The United States, which consumes liquid fuels at an extremely rapid rate, will undoubtedly pass through its peak of domestic petroleum production at a considerably earlier date than will the world as a whole—perhaps as early as 1965-70. As we approach the depletion of our domestic reserves, a very real question will arise concerning the extent to which we permit ourselves to become dependent upon imports of petroleum from abroad." The intervening twenty years have shown the disconcerting clarity of their vision, marred only by their unjustified optimism that the problem would have been recognized in time, for they continue: "In any event, however, it seems highly probable that by the end of the century we will be hydrogenating coal[2] on a very large scale." Twenty years of inaction have made this outcome now highly *improbable* without massive national effort: as we will see in detail in Chapter 7, we have not even begun to produce synthetic fuels to replace those whose decline Harrison Brown and his co-authors saw so clearly.

---

[1] Harrison Brown, James Bonner, and John Weir, *The Next Hundred Years* (New York: Viking Press, 1957).

[2] To produce synthetic liquid and gaseous fuels from coal.

Their words were largely unheeded—perhaps one hundred years is a span of time too great to concern us. More immediate and eloquent were the salesmen of ever heavier and more powerful automobiles, of bigger and better appliances, more strident were the proponents and opponents of nuclear power generation. For too many years, the government itself was misled by ever-escalating estimates of our petroleum resources from its own Geological Survey. When, finally, Richard Nixon announced his ill-considered "Project Independence" with the aim of meeting by 1980 America's energy needs from America's own energy resources, we were simply not ready to believe him, nor did we wish to. Subsequent events in that presidency turned an impossible goal into a half-remembered farce. There is still a massive credibility gap. President Carter's National Energy Plan,[3] while emphasizing conservation, still dreams that in 1985 the total amount of energy from domestic oil and gas will equal that available in 1977. Where is it to come from?

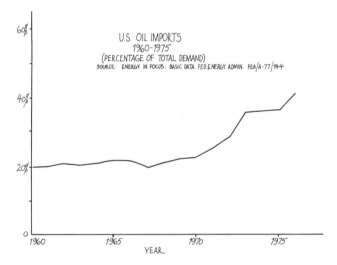

The energy problem is upon us. We have seen that our own production of oil has dropped to three billion barrels a year and is declining. Our national appetite for oil presently amounts to some six

---

[3] Executive Office of the President, Energy Policy and Planning, *The National Energy Plan* (Washington, D.C.: Government Printing Office, 1977).

billion barrels a year. Imports are climbing . . . and are costing us our shirts.

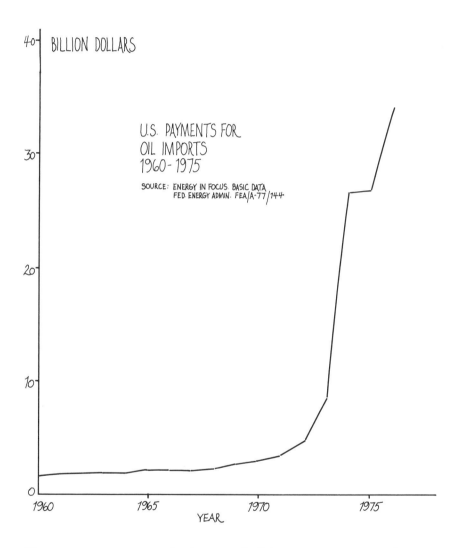

The overseas payments for imported oil in 1978 were equivalent to a remittance of over $150 for every man, woman and child in the United States.

Although conservation alone is not the answer, it is clearly an important first step. The most visible of the energy wasters are of course the heavy, powerful, and luxurious automobiles, the Lincolns, the Cadillacs, and the Toronados into which we have poured so much gasoline over the years. The massive V-8 engines with their vast excessive power have drained away the rich legacy of ages past. We still love these symbols of opulence and power, but is our passion almost over? Perhaps in future years, to drive one may be as antisocial an act as tipping a can of trash on the street. Nevertheless, if you have one, maybe you should not junk it or sell it. Put it away and keep it safely garaged for twenty years or so. You may then have a real museum item, a kind of latter day Edsel that another generation of children can gape at, gleeful and amazed that people could ever have felt the need for such an inefficient, uneconomical monster to carry them about.

Automobiles presently consume some 15 percent of our total energy usage, about the same as all other forms of transportation put together—trucks, buses, aircraft, railroads and ships.[4] As much as 41 percent is used in industry, the hidden uses, in the refining of steel, aluminum, and other raw materials, the processing, molding, shaping, and joining of them into countless things, television sets, pens, railroad cars, chairs, books, feedstocks, aspirins, light bulbs, airplanes, fertilizers and, of course, those cars. The rest is consumed, at roughly equal rates, in our homes on the one hand and in our offices, shops, schools, and hospitals on the other. Energy for space heating is the biggest single category both in residential and commercial establishments, amounting in all to about 18 percent of our energy supply.

Here then is another dimension of the precarious energy balance. The energy use of which we are acutely conscious is the energy that we pay for directly, as gasoline for our cars and in our monthly utility bills. Yet this amounts to about only one-third of the national energy bill. We pay for it all, one way or another, but the rest is hidden from us in the prices of food and clothing and almost everything else on which we spend our money. The need for energy and the use of it is ubiquitous, it permeates almost every activity, every enterprise. Seventy-five percent of it all comes from oil and natural gas, those finite and devastatingly short-lived non-renewable resources.

The energy crunch is not an incremental problem, a small add-on problem like a minor overdraft at the bank. It is a 75 percent problem and affects almost every phase of our lives. This is the magnitude of the revolution with which we must cope.

In addition to energy conservation, we clearly need alternative sources of power, massive sources to replace the energy supply that is declining. We need new sources of megapower, big power, power on the scale of the billions of barrels of oil that are disappearing, from completely new sources that we have not yet begun to tap. Megapower is the power needed to run a whole city, not just the houses but the factories, the schools, and shops, to light the streets and propel the buses. The problem of the missing energy source is a megapower problem.

At the other end of the energy scale is what we might call micropower—power on a scale that we are personally familiar with. It is

[4] J. M. Fowler, *Energy and the Environment* (New York: McGraw-Hill, 1975).

the power that we individually use and pay for directly. Micropower is the power needed to heat our homes, drive a dishwasher or an air conditioner. A single automobile consumes micropower; so does a gasoline lawnmower, a wood stove, and a domestic hot water system. An air conditioner uses more power than an electric fan does, but it is all micropower, power on a domestic scale. A roof-top solar collector represents micropower—a nuclear generating plant is megapower. An ordinary windmill is micropower, a hundred tankers filled with oil carry megapower.

The consumption of micropower adds up to megapower because there are so many of us. A million electric toasters, turned on at the same time, will make demands that not even the largest nuclear power plant can satisfy. In the same way, the saving of micropower by conservation or by substitution adds up to savings of megapower, and this, in turn, means oil that we do not have to import or nuclear wastes that we do not have to dispose of.

The individual elements of micropower may be small on a national scale, but they are substantial in terms of our personal budgets. It may presently cost $1,000 a year to heat a modest home in winter and to cool it in summer, and another $400 a year for gasoline to support only an average amount of driving. In a few years, the bill may be twice as much for the same thing. Individual savings in micropower may help the nation a little bit (and every little bit helps), but more directly and immediately, it will deflect some of the increased costs that we would otherwise have to bear. It is quite obvious that the more micropower we can generate or save ourselves, the less vulnerable we will be to the inevitable escalations in the prices of oil and natural gas as the supplies dwindle.

Micropower consumption adds up to megapower; could the generation of micropower also become megapower? Suppose we drove only small cars, suppose we all had solar collectors on our roofs and windmills in the backyard connected to electric generators—how much would that answer our needs, both our personal needs and our national needs? Could we achieve our own "Project Independence" that way? It would be pleasant to think so, but alas, this is little more than a dream. In terms of *personal savings*, yes; it may make a great deal of sense, but in terms of a *national solution*, it is only a beginning. We would still need power to drive our industry, our factories and steel mills, to transport us about even in high-efficiency cars. But there is yet another reason why each of us is unlikely to achieve our own energy independence: the reason is very simple.

Look back again to the beginnings of human culture. In what we are pleased to call a primitive society, tasks were shared more equally than they are today—everyone gathered food, helped with the building of shelter, and all the able-bodied men joined in the fighting. As groups became larger, more complex and more settled, specialization occurred. Some who liked it or were better at it would do the building for all, others would hunt or grow food for all, others again would do the fighting. An increasingly industrialized society required further specializations; today it is so complex that almost no task is carried out each for himself or herself. Not even personal grooming—we still do most of it ourselves but our hair is generally cut by professionals. A doctor looks after the medical needs of many people, not just himself, while at the same time he may be a very poor carpenter and a hopeless accountant. Others of us are preachers or lawyers, undertakers or electricians, teachers or trashmen. Most of us work at a particular trade or profession presumably because we like it; perhaps we are even good at it. Some of us (including doctors) may not much enjoy the work we do or be very good at it, but we do that particular task for many other people, both because it pays and because it enables us to do the additional things that we would prefer to do. Not everyone has the space for a backyard vegetable plot and the inclination to look after it, and not everyone is going to be able to install, or be bothered with, solar collectors or windmills. Perhaps this is an unduly pessimistic outlook, but for most of us, there are many other things we would rather do with our lives and our spare time—watch baseball, go sailing, talk with friends, write books, play tennis, or just goof off. Being in the part-time power generation business may be, for many, another set of aggravations and another collection of things about the house that can break down. If the technology is ultimately developed to the point where the demands of micropower generation on our time at home are reduced to the levels we presently accept in changing fuses or light bulbs, or even in servicing the car, perhaps many of us will do it, but the technology has still some way to go.

Unfortunately, it is fairly safe to assume that not everyone will generate all or even a part of his own energy needs. Moreover, the missing power for industry and commerce will not be generated by a myriad micropower sources. Megapower problems demand megapower solutions.

Nevertheless, there are a number of important qualifications that we must remember. First, the conservation of micropower does add up to megapower, and this *is* possible for us all. It is relatively simple and painless and we will discuss easy ways to do it later, but whether we like it or not, conservation will be forced upon us by the ever-increasing costs. If we let it, energy will simply become a larger and larger proportion of our personal expenditures, and to have money available to spend as we wish, we will begin to give more than a moment's thought to our energy expenditures as we do in our food expenditures before choosing between hamburger and sirloin. Conservation will save us money.

Secondly, for those who care to generate some micropower for their own use, the savings can be great. Solar collectors to provide hot water and domestic heating may be marginal in terms of present costs for many areas of the country, but it is almost inevitable that the balance of costs will shift. The prices of gas and oil will almost certainly rise faster than the costs of construction, installation, and maintenance of solar systems and, in the long run, the people who have them will be better off. They will still pay for the hidden parts of the national energy bill as do we all, but at least their direct and personal consumption will be checked or reduced.

Another aspect of the pattern of energy use concerns the different qualities of energy needed for different purposes. Electricity represents high grade energy; it can be used to power motors, stereo systems, and computers. The energy content of gasoline is high-grade energy. Hot water from a domestic solar collector is, on the other hand, relatively low-grade energy, yet ideally suited for heating the home. The most effective pattern of energy use (and the cheapest) involves matching the quality of energy to the use that is to be made of it. It is an increasingly expensive waste to consume high-grade energy for purposes such as space heating and domestic hot water when alternative low-grade sources are available. The generation of micropower, generally in the form of low-grade energy, will permit more of our high-grade energy supply to be used for purposes which it alone can satisfy.

Finally, one might reasonably ask which of the micropower sources, the ones that we can install and use for ourselves, can be scaled up to become megapower sources. Remember that the expansion in size is formidable—it is the comparison between a roof-top solar collector and a thousand acres of them, hard against one another. Maybe solar systems can be scaled up, and this is one of our collective options; new industries will be required, and these can hardly fail to help our economy. Other schemes, advocated enthusiastically by some, appear far more doubtful. An ordinary windmill represents micropower, and moreover, it works only intermittently, when the wind blows. Windmills are very useful for pumping water from a well to a tank, where it can be stored for use on quiet days. In isolated regions, a windmill can generate electric power to re-charge batteries for domestic electricity, but present day batteries are extremely expensive in terms of the electricity they can store. An air conditioner would exhaust an automobile battery in fifteen minutes. With research, ingenuity, and inventiveness, a technical breakthrough, an economical solution to the problem of storage of large amounts of electrical energy may yet be found, but this development has not yet arrived. To solve immediate problems, let us not count on break-throughs not yet made. Some people dream about systems of huge windmills that could generate substantial amounts of electrical power, sited along the windy coasts of New England. The National Aeronautics and Space Administration has dreamed to the extent of a one million dollar project for construction of a windmill in Ohio, with sixty-two foot blades, each weighing one ton, mounted on a one hundred foot tower. It works after a fashion, but has been plagued by design and vibration problems—in its first nine months it operated for only thirty hours. In any event, the power produced by so large a structure, even when it is working, is still little more than micropower, one hundred kilowatts compared with the steady flow of one million kilowatts of a nuclear generating plant.

Amory Lovins, writing in the magazine *Foreign Affairs*, October 1976, divides possible energy strategies into a "hard path," essentially our present one, pushing harder and harder on all our remaining resources of oil, gas, coal and uranium, and an alternative "soft path," essential elements of which are conservation and the development of new "soft technologies" for the generation of energy. Soft technologies operate on renewable energy flows, the sun, the wind, the forests, and wastes of various kinds; they are conceived as environmentally benign, dispersed, smaller scale systems, with the energy outputs matched to energy uses.

There is little question that we cannot continue on our present "hard path"—the future resources of oil and natural gas in particular are not there to support it much longer. Yet the alternatives, the soft technologies, are still only sources of micropower and many of them work only intermittently. They can already be useful in supplementing our individual energy consumption, but heavy industry demands megapower no matter where it comes from, and the soft technologies are not yet ready to supply it. They capture a part of the renewable energy which flows through our earth system. Which of them can be scaled up from micropower to megapower? Will they still remain environmentally benign? The urgency of the questions is such that the answers to them can hardly be deferred much longer. We need *both* large power sources and smaller, widely dispersed ones. It is not a question of one to the exclusion of the other. The only questions are how rapidly the soft micropower sources can be harnessed, and how rapidly some of them can be transformed to megapower to replace our present dependence on the dwindling base of oil and natural gas.

Megapower and micropower—these are the two scales of energy production and consumption. Micropower is what we personally can do something about. Megapower demands collective choices, collective action. Micropower savings add up to megapower, but we still need megapower to drive our industry, and to keep our country going. We will never power a steel plant with windmills.

6
Personal Choices: A Survival Guide for the Coming Crunch

By now, it must be clear that we are faced with a rapidly contracting base of our dominant, non-renewable energy resources, and with the disagreeable but inescapable prospect of a continuing escalation in energy costs. The megapower problem can be solved only by the collective resolve and collective action of government, industry, and people. On the other hand, the micropower problems, which impact so heavily on our personal budgets, are on a scale that we individually can do something about.

One reaction to the dismal scenario might be: To hell with it all! To hell with our children and theirs! Let us find ourselves a little place where we can grow our own food, re-cycle the wastes as compost, cut our own wood as fuel, pump our own water with a windmill, go to bed when it's dark, and rise with the dawn. The world can go its own hectic, foolish way, and we will go ours in a kind of Thoreauesque tranquility.

Perhaps.

There is, indeed, an elemental satisfaction in the *notion* of living our own lives with our own resources, of being dependent on no person, and on no outside society for our needs. The reality, however, may offer less satisfaction than the idea. We would certainly reduce (but not eliminate) our individual dependence on outside energy sources, but unfortunately our taxes must still be paid, and our tools, even simple ones, are beyond the capability of most of us to make for ourselves. But more fundamental, I think, would be the dissatisfaction of a state in which the mere sustenance of life is the main business of living. It was this discontent, together with the need for protection, that brought man together in communities so long ago, and ultimately made possible the many real accomplishments in the arts, sciences, literature, and technology, of the past five thousand years (together with, of course, the less desirable accoutrements that came with them). A return to the peasant society of even the Middle Ages, with its confinements, rigors, and brutally heavy labor would have few lasting attractions for most of us.

Instead, let us start from where we are now, with all the constraints, imperfections, problems—and opportunities. We must expect that the prices of the primary energy sources, oil and natural gas, will continue to rise, and that this will mean immediate and direct increases in the costs of heating, cooking, and gasoline for the car. It will also lead to increases in the costs of electricity, our secondary energy source, since many electricity generating plants are powered by gas and oil. To be sure, there are strong incentives for the utility companies to switch back to coal for electricity generation, but for many reasons, some of which we will consider later, that will not offer much price relief. There is an important thing to remember about electricity: In terms of heat supplied, it is at least three times more expensive than are the primary fuels, gas and oil, used directly. The reason is that, to generate most of our electricity, these fuels are burned at the generating plant, producing steam in the boilers that drive turbines attached to electric generators. The conversion of chemical energy (in the oil, coal, or natural gas) through heat to electrical energy is a process that is always very inefficient—the output of electrical energy is only about one-third the input of heat energy. This is not the result of incompetent engineering or wastefulness on the part of the utility companies, but is an inexorable law of nature, one of the laws of thermodynamics that governs the conversion of heat to other forms of energy. There is a maximum possible efficiency of this conversion; it is quite low and modern power plants operate close to the best possible limit. Anyway, to generate one unit of electrical energy, the utility company must burn about three units of heat energy from gas, oil, coal, or uranium. If we are contemplating using that electricity simply for heating or cooking, we are generally at least three times better off to use the primary source ourselves rather than having the utility company use it less efficiently in converting energy from heat to electricity, which we convert back to heat. Electricity can, of course, do things that the other sources cannot—drive electric motors that are clean, compact and efficient, and these, in turn, power refrigerators, sewing machines, and sump pumps. We do not have gas TV's yet, nor are they likely.

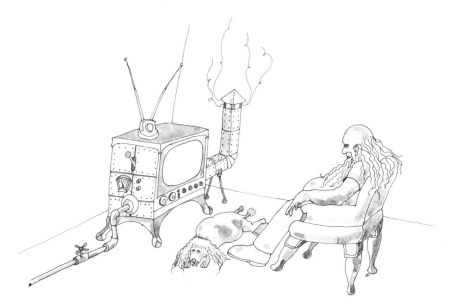

The first line of defense against the prospective cost increases is, of course, energy conservation, and this is largely a matter of perspective interpreted with common sense. Conservation is rather prosaic and undramatic, but it can make a surprising difference. Suppose, for the sake of a concrete example, that you have a house in the Baltimore-Washington area for which the direct domestic energy costs are presently $1,500 per year. The climate there is, in a sense, close to the U.S. norm. It has neither the extremes experienced in, say, Minnesota nor the general mildness of Florida; the reader from those regions can make appropriate adjustments from this norm. Of the $1,500 per year, some $1000 will characteristically be used to heat the house in winter, $200 will be needed to supply hot water, $70 will be used in cooking and the remaining $230 for refrigeration, air conditioning, light, and so forth. Minnesotans will spend more for heating, Floridians less on heat but more for air conditioning.

The biggest part of the bill is for heating, and here are the biggest opportunities for savings. Look around the house and in the attic. If the uppermost ceiling is not insulated, more than half of your $1000 of heating is flowing upwards, through the ceiling and the roof, heating not you but the neighborhood.

Six to eight inches of insulation above the ceiling would probably cost a few hundred dollars but you will recover the investment in a year or two, and from that point, you are ahead all the way. It will cost less if you do it yourself—it is a rather tedious job but quite easy. (Remember not to step on the ceiling between the joists or you may fall through.) Adequate ceiling insulation can save as much as 40 percent of a heating bill in a house previously uninsulated, or almost $400 each year even at present costs. Most older houses and many newer ones, even if they are insulated, have an amount that is inadequate for the present and impending balance of costs.[1] If you already have two to four inches of insulation in the roof, another four inches will save you something like $150 a year.

The insulation of walls in an existing home is more difficult and more expensive, usually a job for professionals. Savings are generally less, but may be worthwhile. Most of the heat loss through the sides of a house occurs through the doors and windows. In an older house, the window sashes or dormers are often loose and warm air flows out as cold air pours in. Put your hand just above the windowsill on a cold morning. Even if the windows are tight, the heat loss through single glass panes is a constant, unnecessary drain. Stopping up leaks, the installation of storm windows and doors together with weather-stripping wherever possible will save another $150 a year or so. Again, your investment will probably be recovered in just a few years. These are big savings—they do require some outlay but the cost is quickly recovered. Even if you must borrow the money to do it, the payments may well be less than what you will save.

Another saving, and an immediate one that costs nothing, is simply to turn the thermostat down in winter. It has been said many times before, it is common sense, and it is true. In the average house, the annual savings are something like $20 per degree that you turn the thermostat back. From 75F to 68F represents about $150 a year. Why stop there? A thermostat set at about 60F will save another $150 a year. One becomes accustomed fairly quickly to a cool house —60F is quite warm enough for most people while active in the

---

[1] The provision of insulation in buildings, domestic or commercial, has not in many areas been covered by any building code. Insulation is not visible and some builders tend to skimp on it.

house, though while sitting or resting one would probably need a sweater. But you can still buy quite a few sweaters for $150 per year.

Summer air conditioning is fairly expensive in our "typical" house, though concentrated in only a month or two of the year. It costs more in Nevada, perhaps nothing at all in Maine. Insulation and a tight house will reduce the cost here too. Some of the excess summer heat enters the house by radiation down from a hot roof; this can be almost eliminated by tacking heavy, shiny aluminum foil to the rafters under the roof to reflect the radiant heat away from the attic.

When installing or replacing an air conditioner, it is well to know that the efficiency, in terms of cooling per power consumed, varies considerably among different models of similar capacity. Of course, the more efficient models are often the more expensive ones, but the savings continue over the lifetime of use. Before installing an air conditioner, it may be worthwhile to ask yourself: "Is it really necessary?" The answer may well be yes, and if so, go ahead. But even if summers are uncomfortable in the house, do not assume that air conditioning is necessarily the answer. Summer discomfort, particularly in the southern and eastern seaboards, is often the result more of excessive humidity than excessive temperature. A temperature of 80° or 85° is quite pleasant if the air is fairly dry, as residents of Arizona and Southern California would proclaim with enthusiasm. Moisture can be removed from the air by a dehumidifier rather than by a full air conditioner. A dehumidifier is generally cheaper to buy in the first place and the power requirements are less—it is cheaper to run. One practical problem, however, is that it may need a drain to get rid of the water condensed from the air.

Except in unusually long hot spells, quite good summer temperature control in a well-insulated house can often be achieved without much energy expenditure at all. You need a window exhaust fan in an upstairs room and an indoor-outdoor thermometer. Drawn curtains or shades on windows with a southerly and westerly exposure will reflect much of the sun's heat throughout the day and windows closed, or only slightly open, will keep much of the hot air outside. In the evening, when the air temperature outside drops to that inside the house, turn on the exhaust fan and open one or two windows at the other end of the house to draw cooler air inside until it is time to go to bed. Think about the flow of air through the house so that it passes through the rooms that you wish to cool. You can, if you desire, leave the fan running all night with an upstairs window open to allow a flow through the house; a fan uses very little power, about a tenth that of an air conditioner. In the morning, when the outside temperature begins to rise, turn off the fan and close the windows—the house will remain cool all day. Does it sound a real bother? In fact, once you have worked out the system, the temperature control of your house takes about as much time as cleaning your teeth.

These very ordinary, simple, and common sense steps can make really big savings in micropower and translate into substantial savings in our budgets. In many houses, a $1500 expenditure for domestic energy can be cut almost in half—*if* the cost of energy remains constant. Being more realistic, one must admit that we may not end up with money in the bank. If, in the next few years, the cost doubles, we will still be about where we are now, but very much better off than we would have been had we done nothing. The savings in micropower also add up to megapower, not enormous megapower, but still significant. Remember that power for residential heating and cooling amounts to about 12 percent of our total national energy usage—reduce that by one third, and we have saved the equivalent of about half a billion barrels of oil a year.

This may be all very well for those who live in their own houses, but what about the millions of us who have apartments or rent houses? We may pay for the utilities directly or we may pay indirectly, as part of the rent. In any event, we may well be unwilling to install storm windows and, in a rented house or garden apartment, to insulate the roof. Some form of collective action among the tenants may be necessary to persuade the landlord that it is in his best interests ("to keep the cost of his apartments competitive" is a better way of saying it) to deflect as much of the increasing energy cost as possible by adequate insulation and storm windows. When the cost of heating is included in the rent, there is little individual incentive to turn the thermostat down; if you are seriously interested in avoiding rent increases triggered by increasing energy costs, it might be advisable to avoid such apartments.

There is often not much thermal insulation between apartments. One winter my family lived in one of the row houses for which Baltimore is noted. Our thermostat was set at about 65F, partly because we liked a cool house and partly because we did not have much money. Our neighbors on either side liked very warm houses; they had their thermostats over 75F. It had been a reasonably hard winter but, a little to my surprise, our furnace scarcely went on at all. In the spring, at a neighborhood get-together, my neighbors on either side were complaining bitterly about their enormous heating bills. I suddenly realized that *their* heat had been flowing through the walls on either side to keep *our* house warm as well—I gulped and held my tongue. If, in an apartment, you pay for your heat directly, you too can win by turning down your thermostat. Your neighbors who are less conservation minded will help to keep you warm.

A fireplace in the house is one of the family luxuries that many of us enjoy so much. Toasting marshmallows on a Sunday night or just gazing into the embers makes for memories that linger. A warm fire on a snowy night draws a family together as TV never will. Yet make sure that the fireplace has a damper to close off the flue when not in use, otherwise the chimney will go on night and day exhausting not smoke but warm air that you cannot afford to lose. Sometimes a blazing fire can, in fact, cause the loss of more heat than it provides. The fire draws warm air from the room and the higher the thermostat is set, the greater the heat loss. It is much better to draw the air for the fire up through the ash-pit, either from outside or from the basement, in order to reduce the warm air loss as much as possible. If no air at all comes from the room, however, the fire will smoke, and this can be a problem in a really tight house.[2] You may have to open a window somewhere, just a little bit, to stop the fire from smoking, and you can feel for yourself the cold air drawn in. Though many of them are remarkably ugly, new high efficiency tubular grates are available commercially, and they can multiply substantially the heat output from the fireplace. So if you are having a fire, turn the thermostat down, allow a little air into the house, then sit down and enjoy it.

There are many ways of making little savings that add up. As

---

[2] A badly designed flue will also make for a smoky fireplace.

parents have told their children for at least thirty years: Turn off the television when you are not watching, turn off the lights when you leave the room. In an unused room or a storage room, turn off the heat if you can and keep the door closed. A five-minute shower uses one third the hot water of a full bath. In the kitchen, use a small toaster oven whenever possible rather than the main oven, particularly if the latter is electric. If you have a gas oven, there may not be much to it either way—electricity for the toaster oven is more expensive but heats a smaller volume. If your dishwasher has a "wash only" cycle, use this rather than the 'wash-dry' cycle. After washing, the plates and cutlery are hot enough usually to dry themselves without the additional heating and blowing in the drying cycle. They may take a little longer to dry, but often there is no hurry.

Some possible energy savings may be more marginal in terms of the extra work they entail. A clothes dryer will labor long and hard to dry a heavy quilt or blankets; if you have a few to do, you may save a dollar or so by hanging them outside on the line to dry. On the other hand, light objects like underwear, shirts and blouses dry very quickly in a clothes dryer, while to hang them outside entails sorting out and pegging up a large number of small items. The bother of hanging them out and bringing them in may not be worth the small savings. In any event, do not use a dishwasher, a clothes washing machine, or a dryer until there is a full load—not an overload, though, or the machine will not work properly.

Frost-free refrigerators use about twice as much power as the ordinary models that must be defrosted periodically. Yet defrosting a refrigerator is one of the messier and less pleasant of kitchen tasks, and it may be worth the five or ten dollars a year to avoid it. The vote on this should be cast by the person who does the defrosting.

Then there are the frivolities, the electric can opener and the electric tooth brush, but they use so little power that it hardly makes any difference. The energy required to build them in the first place is likely to be larger than the energy they will use during their whole lifetime.

When the time comes to remodel or to replace appliances, the stage can be set for future energy conservation. In cooking and heating, reduce electricity to the minimum—you will be better off with oil or gas heating and gas cooking. Make sure your new gas stove is pilotless, with a device like a spark plug for lighting the gas. A pilot light, burning constantly, can use as much gas as is needed for the actual cooking. Fluorescent lights give much more illumination per watt (per energy consumed) than incandescent (filament) bulbs. Think about replacing the kitchen exhaust fan by one that filters the air through activated charcoal. In winter, an exhaust fan draws out not only smells and smoke but warm air as well, and cold air comes in somewhere else. If you must replace the hot water heater, try to relocate it as close as possible to where the hot water will be used. In many houses, one must run the basin faucet for some time before the water is warm, yet all that water was heated in the first place and has cooled while standing in the pipes. It is not really economical to insulate the hot water pipes since the water often stands there long enough to cool anyway in spite of insulation, but if the water heater is closer to the kitchen and the bathroom, the wastage will be less. On the other hand, it is often worthwhile to add further insulation to the hot water tank itself. When replacing a hot water heater or a furnace, pay great attention to the relative efficiencies of different models you may consider. Since heating is so large a fraction of the total bill, even a few percent reduction in the heat that goes up the chimney is worthwhile.

Some of us are inadvertent energy guzzlers. An ornamental outdoor gas light adds rustic charm to a garden and may give a sense of security—yet it consumes about $100 worth of gas a year, some 10 percent of the total heating bill! If you need the light, replace it with an electric one. It costs very much less per hour and you can switch it off during the day.

Finally, a few are fortunate enough to have an outdoor swimming pool. We may like it to be reasonably warm, but heating will cost a small fortune. If you can afford the pool, perhaps the cost does not matter to you, *but the energy matters to us all*. Practically all of the heat loss from a swimming pool, even on a cold day, is the result of evaporation. As liquid water evaporates, it absorbs energy, large amounts of it, and this is drawn from the rest of the water remaining in the pool. It takes just about as much energy to evaporate a gallon of water from the pool as it does to boil dry a gallon of water on the stove. This is the heat you would have to supply to keep the pool warm. Evaporative losses can be reduced greatly by covering the pool when not in use, preferably by a transparent material that allows the sun's rays to penetrate while preventing the evaporation. A cover like this is a kind of solar collector—it will warm up the pool much earlier in the spring but in high summer, you may be obliged to leave it off to prevent the water from becoming unpleasantly tepid.

Other opportunities for conservation may become feasible within a few years. For at least twenty years, a heat pump has been advocated as an energy saving method of heating in winter, cooling in summer. A heat pump works rather like a refrigerator; in summer it cools the house like an air conditioner and releases the heat outside;[3] in winter, the cycle is reversed and the heat is pumped inside. A heat pump *will* provide three or four times as much heat per unit of electricity consumed than one would obtain by direct electrical heating. However, because of the losses in generating the electricity in the first place, the heating per unit primary fuel consumed is about the same as if you heated directly with gas or oil. The energy costs are then comparable. Moreover, the initial capital cost of the heat pump is high and so far the maintenance records have not generally been good. The technology is one whose time has perhaps not yet come, but when we no longer generate so much electricity by burning fossil fuels, the balance may shift.

With a little ingenuity and common sense, almost everyone can find opportunities for energy savings around the house or apartment in ways that will not make life unpleasant, tedious, or laborious. The individual incentive is the ability to deflect increases in energy prices —no doubt we can find better things to do with those dollars than let them float out the windows or up through the roof.

The other big item of personal energy consumption is, of course, gasoline for the car. A gallon of gas saved translates directly into oil that does not have to be imported and into dollars that do not drain towards the Middle East. It is certainly realistic to anticipate steadily increasing taxation on gasoline in an attempt to reduce the currency drain, so that the price is likely to rise even faster than will other

[3] Reach around the back of your refrigerator when it is running and you can feel the heat being released.

energy sources, some of which can be replaced more readily by domestic coal. In personal terms, when buying a car, we must anticipate filling it with gasoline at $2.00 or so a gallon. We may enjoy the luxury of a big and powerful car, but the pleasure will be somewhat diluted by pumping in $40 worth of gas for four hours' driving. If we run the air conditioner incessantly, it will cost even more.

The case for choosing a gas-stingy car is so obvious that little more need be said.

All this has been concerned with conservation, with reducing the amount of energy we buy from the utility company, the oil supplier or at the gas pump. What about capturing our own energy?

It has become increasingly fashionable to suppose that the warmth of the sun will solve all our energy problems. Many schemes have been proposed, some realistic, some little more than fantasy. Solar energy *can* provide megapower and we will see later a couple of ways in which this can be done. It can also provide micropower for the house, to supplement but probably not to replace entirely the energy we buy from other sources. The idea of roof-top solar collectors is not new, nor are such devices untried. It is estimated that perhaps a million solar water heaters are already in use in Australia, Japan, and Israel.[4] In the United States, by 1972 there were only a few dozen residences and small buildings heated with solar energy. In 1978, there were rather more than ten thousand; the potential number of domestic solar energy installations in the United States must be in the tens of millions.

The simplest and most practical type of solar collector consists of black absorbent panels mounted on a flat roof or one sloping to the south. Ducts or tubes carry air or liquid which, by being heated, remove energy from the panels. Insulation at the back reduces energy loss downwards and glass plates on top (that must be kept clean)

[4] J. A. Duffie and W. A. Beckman, "Solar Heating and Cooling," *Science* 191, 16 January 1976, p. 143.

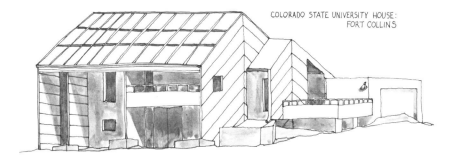
COLORADO STATE UNIVERSITY HOUSE: FORT COLLINS

prevent as much re-radiation as possible. The heat is stored in an insulated tank for water or perhaps in a pebble bed if hot air is used; the stored heat can be used to provide hot water or space heating for the house. It is possible, though less efficient, to use the energy for cooling also, in the manner of a gas refrigerator, but workable and economic units are probably some time away.

The advantages of solar energy are evident—it is a renewable, nonpolluting energy source. But how much could we realistically expect of solar panels in providing domestic heating and hot water throughout the year? One awkward fact is that solar panels will produce more heat in the summer when we need it least and less in the winter, when the demands are greatest. Another is that the heat from the sun is fairly diffuse; one needs to cover a substantial fraction of the roof area of a house to gather enough energy to make a significant impact on the needs throughout the year. Nevertheless, a collector, eighteen feet by forty-five feet in area, even in Wisconsin, can provide more than enough hot water for a medium-sized house between May and October, and cut the heating bill by half during the winter months. If we were able to capture *all* of the incident solar heat on the roof, we would have more than enough throughout the year, but the maximum efficiency is only about 50 percent, and supplementary heating will still be necessary. In summer, there is much more available than we need, but in winter there is a shortfall. Nevertheless, the auxiliary heating system need have a capacity only about half that of the same house without solar collectors, and the total bill for heat and hot water would be slashed by two-thirds.

That sounds fine, but remember that large investments in equipment are needed to make those fuel savings. The essential problem that each householder will have to solve for himself is the decision on when the cost of the extra investment becomes worth the annual fuel savings. If the cost is more than about ten times the annual saving you expect, then it is not presently worth it. But remember the basic problem—those annual fuel costs are going to rise and keep on rising. Even with conservation, we can hold them down only so much. What is marginally worthwhile now may be giving you a handsome profit in ten years' time.

Even if you decide that now is not the time to install your solar collector, you will certainly keep it in mind. Present domestic installations are still rather primitive, though the field is burgeoning, and technological and design improvements are coming rapidly. The next ten or fifteen years will probably see something like the early decades of automobile production, when hundreds of companies made an astonishing variety of gas buggies, steamers, and electric cars, each built one at a time. Many of the companies folded or merged with others and today only a few remain. Solar power will probably go through the same kind of sorting out. In practical terms, you may have trouble ten years from now in obtaining replacement parts and service for some of the early models, unless they are so simple that you can do it for yourself. Nevertheless, as the costs of collectors and other components decrease as a result of mass production (or of our "doing it ourselves"), the economics will improve. The price of fuel will rise, the costs of solar collection systems will decline—the crossover point will come eventually, and it may come quicker than we expect.

Yet domestic solar collectors are not for us all. They are not much use for high-rise apartment dwellers, and it seems that a larger proportion of us may well be living in apartments or condominiums as time goes by. Floridians do not need the heating, but they would like solar cooling and practical systems for that may be some time away. Nevertheless, it is almost inevitable that the installation of domestic solar collectors will become widespread; they are by no means the answer to all our problems, but they will help.

With careful design, a new or re-modelled house can capture much of the energy needed for heating without even the use of roof-top solar collectors. The house should, of course, be well insulated; windows facing north should be either small or eliminated altogether. Those facing south should have double panes and be as large as possible while protected from the summer sun by an overhanging roof. To reduce night and day variations in temperature in winter, heat from the sun's rays can be absorbed by a thick masonry wall during the day to be released naturally to warm the house at night. A number of variants of these "passive solar systems," so called, include a roof pond for capturing and storing the heat and a simple greenhouse with a heat storage wall that can be attached to an existing house. Despite the novelty of these ideas in contemporary American architecture, the collonades and thick masonry walls of the old Spanish colonial style served the same purpose—perhaps we are rediscovering an old idea, but one not less useful for that.

Windmills are much more doubtful as a generally useful micropower source. The cost of construction per unit power obtained is generally a good bit more than solar collectors and windmills work only intermittently. They are very valuable in pumping water from wells to tanks in rural areas and they have long been used for this purpose in the American West and in Australia. One can certainly attach an electric generator to a windmill, but what would you do with the electricity from your own windmill? A windmill in Oklahoma, six feet in diameter, would deliver on average about one kilowatt—about enough to run an electric toaster—but much of the time, it would not deliver anything at all. Some have proposed hooking up the wind generator to tie into the utility grid of the local power company.[5] When the wind does not blow, one could draw from the grid, but when it does, the meter would run backwards. Before you dream about the power company owing *you* money, however, remember the cost of installing and maintaining your windmill. Again, the enthusiasm of the power company about buying your electricity may be qualified—the company must still install and maintain the capacity for power generation when the wind does not blow.

Nevertheless, the cold facts of dollars and cents will encourage us

---

[5] See, for example, the article by Henry S. Reuss, *New York Times*, June 27, 1976.

more and more to look carefully to our energy use, to consume less, and to provide part of our own needs. The profligate age is over. We will batten down the hatches for the coming storm, and wise is the family that is ready in good time. On a national scale, if we could save or provide for ourselves half our present domestic needs, this would amount to about 10 percent of the total energy use. But remember that the overall problem is a 75 percent problem—there is still a long way to go.

7
Finding the Missing Megapower: Immediate Alternatives

The simplest "solution" to our dwindling supply of domestic energy is to import more, at whatever the cost. This is the way we have been drifting, the easy way, but it is the road to national impoverishment, economic decay, and personal hardship.

Whether we like it or not, the prices that we presently pay for all our energy are a reflection of the prices the OPEC countries charge for their oil. As oil prices rise, part of the energy demand is shifted to the ready alternatives, particularly coal, and the price of it rises as well. Gas follows, sooner or later. In the matter of energy costs, the OPEC oil ministers have us by the ear. When, in 1973–74, they tripled the price of oil, we had no real alternative than to pay. Since then, the steady increases have just about kept up with inflation, but we are promised larger ones in the future.

Why *will* the price of oil continue to rise? Sheik Ahmed Zaki Yamani of Saudi Arabia and the other Arab oil czars are not stupid men. Put yourself in their place and the answer will become obvious. They know that our rate of crude oil production is declining and will continue to do so in spite of the Alaskan fields. They know that the world rate of discovery is now near its peak and perhaps already declining; world production, concentrated in the Middle East, will similarly peak and then begin to decline in the grand and inevitable cycle that Hubbert foresaw. The United States is twenty years ahead of them on the cycle (one kind of leadership that we would prefer not to have) and their own oil bonanza has only twenty or thirty years to go. It will be over in one generation.

Except for oil, these countries are desperately poor. The rock and scorching sands of the desert offer little refuge. Nature is cruel and survival is an unrelenting struggle. For more than three thousand years in places, nomads have crept from one precious oasis to another in search of the water that was life to them and to their miserable flocks. Their culture, which flowered in the graceful architecture of Isfahan and Medina, their literature, and their caligraphy rested on a thin and precarious resource base.

Suddenly came the oil, the black gold that the industrial West could consume with a voracious and apparently insatiable appetite. Huge quantities of oil lay beneath the desert rocks—much more than had ever been in the United States, and about half as much as all the rest of the world put together. It literally gushed out from the sands of Libya, Saudi Arabia, Iraq, and Iran—the incremental cost of producing it was less than a penny a gallon. For the first time in millennia, the outside world was desperate for what the Arabs had, and the outside world would pay for it. Their chance had come, a brief and fleeting chance that will not come again. They have twenty or thirty years to re-build these ancient lands, to buy the technology that alone can transform them, to make the investments that will ensure an income when the oil is gone.

No, they are not foolish men. They have this one opportunity and they have seized it. They must get as much as they can for that black

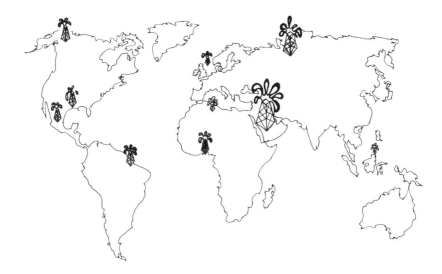

gold—the price of oil must be as high as the traffic will bear. Too great an increase at once would be self-defeating. Crippling deficits incurred by the western nations would lead to massive currency devaluations and a loss of those investments that will be needed when the oil tapers off. The noose must be tight and made even tighter, but not quite enough to strangle entirely.

If you were Sheik Yamani of Saudi Arabia or one of his ministers, proud nationalists all, would you not do the same? The United States is still the richest country in the world—after three thousand years of waiting, *why should they be generous?*

The middle East is not the only place where substantial amounts of oil remain to be extracted. The Soviet Union and China have resources that are not quite as extensive to be sure, but comparable.[1] When at the Geological Institute in Novosibirsk, Siberia, a few years ago, I was shown maps of huge oil provinces, drilled and proven but not in production, stretching for almost a thousand miles east of the Urals. These fields are simply being held in reserve until the time that they are needed—the U.S.S.R. will not face our problem for another forty years or so. Would they sell us oil, in spite of political and

---

[1] M. King Hubbert, "Energy Resources of the Earth," *Scientific American*, 1971.

doctrinal differences? Maybe; they need the foreign exchange, but we would be simpletons to expect a better price than the rate set by the OPEC ministers. The cost would be just as ruinous, and the political ramifications may be unacceptable. The North Sea finds are large, but comparable only with Alaska. They will give Norway and Britain their moments in the sun. Reserves in Latin America, particularly Venezuela and Mexico, are considerable, but no larger than our own. There is a great deal of money to be made from these fields, but they do not begin to compare in extent with those of the Middle East or of the Soviet Union. In terms of present United States consumption, they amount to only a few years' supply. No matter where we look, if we continue to rely on imported oil, we face the prospect of a continually increasing price and an escalation of the already enormous drain of dollars abroad to pay for it.

Nevertheless, we are not quite powerless. We have the ability, if we have the will, to put a lid on those prices, but we will have to be ready to do it when the time comes. There are perhaps 1,000 billion barrels of oil locked up in the shales of the Green River Formation in Wyoming, Utah, and Colorado. The environmental penalty in extracting it may be extreme, and the process itself may cost $20 a barrel, but the oil is there. We can probably obtain almost unlimited power from the oceans at costs which today are already marginally economical—provided we go ahead with the building of ocean thermal plants needed to produce it. We could be in the position where we could tell the OPEC ministers that we do not want their oil at $20 or $30 a barrel, or whatever the asking price may be in 1988. However it will avail us little that these or other alternatives are possible and theoretically economical if we are not already implementing them, if we are not *already producing megapower* on the scale that we need from resources that we control.

The environmental damage involved in the extraction of shale oil may be too great to bear, and the first ocean thermal plant is not yet built. For the immediate future, we must look to the resources that we have, the ones that can be developed rapidly from an existing base, even if they are not the ones that our children will use.

Remember the dimensions of the problem. Our own supplies of oil and natural gas are dwindling to the point where in a very few years they will make an insignificant contribution to our energy supply. Yet as long as we drive automobiles that are anything like the ones we use today, as long as trucks are widely used in hauling food and goods, we will continue to need liquid fuel to power them. Electric cars would depend on technological breakthroughs in battery storage not yet made. We will not stoke our automobiles with pellets of coal or fire up a Stanley Steamer with half a cord of wood. High pressure gas, such as hydrogen, used to fuel a car could make the smallest accident a disaster. There seems to be little alternative—we will need liquid fuel, synthetic gasoline of one kind or another.

We also need gaseous fuel for cooking and that part of domestic heating not provided by solar collectors. We will not easily give up the advantages that made natural gas a premium fuel—the clean burning, the ease of overland distribution and the convenience of use. But this gaseous fuel will necessarily be something other than natural gas. We need electric power for lighting, for machines and electronic gadgets but it is not necessary to burn precious liquid and gaseous fuels to generate it. The simplest substitution is coal and conversion back to coal is being forced upon the utility companies already. Nuclear power plays a part in electricity generation, though its ultimate contribution may be less than was once envisioned. Hydroelectric power supplies 16 percent of our electrical needs, less than 4 percent of the total energy supply; the opportunities for expansion are limited. Other sources are still largely in the future. We will come back to them in the next chapter.

Conservation in our personal energy expenditures will save us money, as we have already seen. Personal energy conservation will also reduce our national need for megapower, but not nearly enough. What are the national opportunities for energy conservation, either by the reduction of total demand or by re-distribution, so that short-lived resources like oil and natural gas are replaced by coal?

Private automobiles presently consume some 15 percent of our total energy supply. The increasing fuel efficiency standards already in place may gradually reduce this percentage as older, less efficient automobiles leave the road. Even so, nearly half[2] of the gasoline consumed by automobiles is used in driving around our cities. Urban mass transport is more than twice as efficient in terms of pas-

---

[2] J. M. Fowler, *Energy and the Environment* (New York: McGraw-Hill, 1975).

ENERGY CONSERVING VEHICLE

senger miles traveled per unit energy consumption. Could it not be made more attractive to catch the bus than to drive our car? Ideally, perhaps yes, but let us be realistic. Unlike the older cities of Europe, our towns grew during the profligate age. The compact city centers decayed and suburbs diffused for miles. Low-density housing developments drew off the more affluent from the crowded city areas, and the new residents of the suburbs could drive where they wanted to go. The neighborhood grocery store might be five miles away but it did not matter as long as gasoline was cheap and plentiful. Now that gasoline is becoming less and less cheap, the dispersal *is* beginning to matter but, alas, we are stuck with it. If we are obliged to use two gallons of gasoline a day to drive to work, a price of $2.00 a gallon represents, capitalized, a penalty of about $10,000 for living in the suburbs. The increasing price of gas may have some uncomfortable effects on the value of suburban real estate!

Public transport *is* attractive if you can catch the bus from the end of the block to where you want to go, and if buses come along every five or ten minutes. It is a hardier soul who will habitually walk a mile to the main road before he can catch the bus, with the prospect of waiting for a transfer and a similar hike at the other end. Unfortunately, the dispersal of our cities is a relic of the profligate age with which we must live for at least some time. It provides a challenge to our ingenuity that we have hardly begun to tackle. No ordinary bus line can afford to provide service through miles of suburban streets in order to collect the odd passenger or two. There are, in places, parking areas in the suburbs where buses or rail transport can be taken for the city, but even these satisfy little of the overall need. We would require a web of fast, frequent public transport lines, rather than just spokes radiating from the city center, if we are to catch the bus, not only to go to work, but to visit Aunt Sarah, to go to the movies, or to pick up a pound of nails. Perhaps we cannot now afford it. On the other hand, before long we may find that we cannot afford to be without it.

Twenty-five years ago, almost 20 percent of urban passengers were carried by bus; today the figure is less than 5 percent. There were good reasons for the shift, and there are now good reasons for shifting back. Whether or not we choose to leave our automobiles in the garage when we go to work depends on how much it costs to run those automobiles, on the attractiveness of the alternatives, and on our willingness to find a small amount of extra time.

For inter-city transport, a train is more than twice as economical (in terms of passenger miles per unit energy consumed) as even the most gas efficient automobile carrying two people. Moreover, if electrified, the train need consume no imported oil. Yet the fact is that our passenger railroads are bankrupt, and the tracks that have not been abandoned are often in woeful condition. Nevertheless, there is hope. Already, in traveling from Washington to New York, it makes much more sense to catch the train than either to drive or to fly. "Metroliners" leave on the hour for the 180-minute journey; to drive is slower (what do you do with the car when you get to New York?), and flying—more expensive—is no quicker in time if you include the struggles to and from the two airports. The traffic on this rail line is already heavy but only a fraction of what it *might* be (overall, about 85 percent of inter-city transport is still by automobile, 2 or 3 percent by rail).[3] Providing this kind of service in other intensely traveled

[3] Ibid.

corridors will require substantial capital investments, but as the cost of gasoline rises, the national savings will become significant.

Industrial and commercial energy conservation can take many forms. Fuel for commercial space heating and air conditioning is a big item. Yet this is one of the hidden energy costs that we begrudge but continue to pay. On a hot day, would we prefer to shop in a store in which warm, maybe dehumidified air is stirred by a fan, even if the prices are a little cheaper? Insulation in factories and commercial buildings will certainly become more widespread in an effort to keep costs down and to remain competitive. Solar heating will help to warm shopping centers and schools as well as homes. The opportunities are as many as they are for domestic savings, but vary greatly from one situation to another; prudent executives and managers will become increasingly alert to them.

National energy conservation—the reduction of total megapower demand in industry, commerce, and transport—does offer limited opportunities to reduce the demand for oil and natural gas. Perhaps, at the outside, the energy use might be reduced by about 10 percent with present technology, but a reduction much beyond that would begin to cut into productivity and effectiveness.

Not all of the oil and natural gas used by industry is consumed as fuel. The chemical industry takes one-third of all petroleum products consumed industrially, and 12 percent of the natural gas. Of this amount, about three-quarters is earmarked as raw material for the manufacture of plastics. Barry Commoner in his important book *The Poverty of Power*[4] argues that the use of plastics should be limited to those articles in which it has unique advantages, such as phonograph records, heart valves, and shatter-proof glass, rather than as a replacement for wood, wool, cotton, metal, and clay which can be produced with much less energy and much less capital investment. True, to produce china plates requires only about one-quarter the energy of plastic plates, yet one must admit to being more comfortable with the latter at a picnic. Again, some synthetic fibres do have real advantages that will not be given up lightly. How many women would be prepared to return to cotton stockings and underwear or be

---

[4] Barry Commoner, *The Poverty of Power* (New York: Knopf, 1975).

able to pay for the luxury of silk? Where is the land for growing the cotton to come from, whose food would it displace, and what are the hidden energy costs? Nevertheless, Commoner's point is a cogent one. Disposable plastic products are not degradable; when burned, some of them produce toxic gases. They litter our roads and our beaches and clog our already enormous "landfills" (trash piles). The archaeologists of one thousand years from now (if there be any) may well have a tawdry view of our culture.

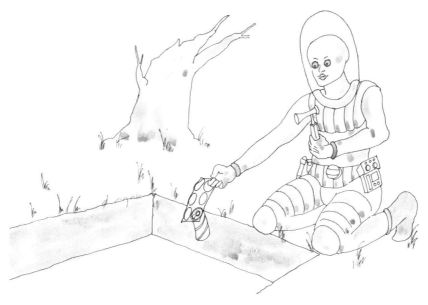

The production of fertilizers also consumes large amounts of energy. Advocates of organic farming claim that it seems to yield about the same economic return as the use of manufactured fertilizers, but with improvements in the workability of the soil and at one-third the expenditure of energy.[5] To a degree, the present excesses in the use of plastics and of manufactured fertilizers may be self-correcting—their cheapness has depended on the cheapness of their energy base. As the cost rises, the balance will swing back at least part of the way.

[5] Ibid.

All of these changes will reduce incrementally the demands for oil and natural gas. More significant effects can be obtained by the substitution of other sources of fuel and raw materials for these diminishing resources. In both the short and the long term, this is the only direction we can go, and the only readily available material to make up the difference is coal.

Our present production of coal is about six hundred million tons per year—some 25 percent of our national energy supply. Present targets call for a doubling of the rate of production by 1985.[6]

Coal mining is one of the dirtiest and most dangerous of industrial occupations. The deep coal miner's working life is threatened by hazards of exploding mine gases, roof falls, and flooding. Even if he survives to retirement, he is apt to be strangled and incapacitated by a lifetime of coal dust inhalation—the crippling disease called black lung. Modern techniques of mining safety *can* reduce these hazards greatly, but the hazards persist. Strip mining is safer for men, but has been devastating to the environment in the coal fields of Appalachia and in the Mid-west. Not only has land been destroyed, but mountain streams, once clear, have been clogged and stripped of life by the yellow sulfurous leachings from the mines. Recent legislation has sought to curb the worst of these environmental abuses; we may begin to halt the further destruction of land for coal, but the forlorn legacy of the past remains.

Coal is a dirty fuel to burn. The clouds of smoke from factories and steam locomotives blackened the English Midlands and our own industrial towns at the beginning of this century. Not only soot and ash goes up the smokestack if we let it, but also oxides of sulfur that damage plants and human lungs. As oil and, later, natural gas became available, it is little wonder that we turned eagerly from coal to

---

[6] Testimony of Mr. J. Schlesinger before the Joint Economic Committee, 1977.

these very much cleaner fuel sources, easier to transport and to use, sometimes even cheaper.

Smoke particles and sulfur dioxide can be removed from flue gases, but the present technology for doing so is astonishingly primitive, having almost stagnated for many years. With natural gas particularly, but also with oil, it was not needed. The most advanced electrostatic smoke precipitators are truly Rube Goldberg devices. Smoke particles are collected on the walls of huge vertical ducts where they agglomerate. Periodically, the walls are literally banged at the side to shake the accumulation loose; some of it falls to the bottom and is collected. Scrubbers to remove sulfur dioxide are cranky and costly, so that most industries are notably unenthusiastic about them, but they can capture up to 90 percent of the emissions. There is a desperate need for new and efficient technology in this area, but it is slow in coming. Many among the coal producing and consuming community are simply unprepared technologically to meet the expectations of society in coping with the pollutants from coal burning. In a coal-burning power plant, the cost of fuel may be considerably less than the cost of cleaning up the emissions, but the price must be paid and it is added to our monthly electricity bills.

Already, about half of our electricity is generated by burning coal, 30 percent uses oil or natural gas, and the remainder is derived from hydroelectric or nuclear plants. The proportions vary from one locality to another, but overall, more than half of the coal we presently produce ends up in electricity generating plants. Some plants which at one time burned coal, were converted to liquid or gaseous fuel, and now, although it is costly, they will be converted back or scrapped. Some of the more modern plants use their fuel directly in gas turbines (devices of the same family as turbo-prop aircraft engines), and these cannot be converted to coal. Nevertheless, if a large fraction of present plants using gas or oil are converted to coal (as is the present plan), our oil imports will be reduced by almost a billion barrels a year (15 percent of our present consumption and almost 30 percent of our imports), and our natural gas consumption by 25 percent. Reduction of oil imports will help stem the stream of dollars abroad; the switch from gas will stretch out what is left of this precious resource. We may end up paying more for our electricity bills, but for the next few years at least, this seems the only way to go.

Gasification of coal is an old technology. The street lamps of Dickens' London were lit by coal gas. Even in recent times, the skyline of the upper reaches of Sydney Harbor in Australia was dominated by the gas works at Mortlake, where colliers berthed two and

three at a time with coal from Newcastle, N.S.W., a hundred miles to the north. In the United States, every city of reasonable size had its gas works for industry and for the home, for cooking and heating hot water. Not much was used for heating the house—it was rather too expensive. Coal furnaces were much more popular at the turn of the century, but they needed to be stoked every day and the ashes had to be removed. Later the furnaces were converted to oil, and one onerous household chore vanished.

Then came natural gas, and overnight the gas works, with their ancient, lumbering technology, were dead. The new premium fuel, which had originally been wasted, was cheaper than manufactured gas and had a higher heat content. The gas works were dismantled and natural gas flooded in, along the pipelines from the fields of the West and South-west. Today, not a single commercial plant is operating in this country to produce gas from coal.

A glance at the relative costs tells the story. The price of natural gas varies enormously. Interstate gas prices in 1976 were about 51¢ per million Btu;[7] recent rulings raised the price towards $1.75 per million Btu and beyond. Imported liquefied natural gas from Algeria presently costs about $3.15 per million Btu. Synthetic gas from a full-sized commercial facility, with the old, inefficient, Lurgi technology, would cost $4.00 per million Btu. As long as natural gas was plentifully available at 50¢ or so per million Btu, the gas works simply could not compete and they disappeared. However, times change. Coal gas at $4.00 per million Btu is roughly the equivalent of oil costing about $20 a barrel, a price level that we appear destined to reach within a very few years.

The basic technology for gasification, the Lurgi and Koppers-Totzek processes, were developed in Germany in the last century. They are relatively inefficient, converting only about 55 percent of the coal's energy to gas, but still, they represent the commercial state of the art. As Allen Hammond, writing editorially in *Science*,[8] points out "It is not an art in which there is a lot of recent experience in the United States. No Lurgi or Koppers-Totzek plants are now operating in this country and, despite growing shortages of natural gas, none are yet being built. This lack of a commercial coal gasification industry may become a subject of increasing concern as gas shortages accelerate, as they seem certain to do. Moreover, failure to get a few U.S. plants built with existing technology—to establish financial, regulatory, and environmental precedents—may prejudice the success of the research program aimed at developing and commercializing improved gasification techniques."

[7] That is, per unit heat content. The heat content per unit volume is smaller for most coal gas than natural gas, so that comparisons based on volume are misleading. The figures quoted are from A. L. Hammond, *Science* 193, 27 August 1976, p. 750.
[8] Ibid.

There are a few sporadic efforts to devise new processes for synthetic gas that may be more efficient and less expensive than the old methods. Some involve a modification of the Lurgi process while others would produce both gas and oil. One technique that is potentially very promising involves the underground gasification of coal that may be too deep for conventional mining. In recent tests in Wyoming conducted by the Lawrence Livermore Laboratory of the University of California, two holes were drilled into a coal seam of the Powder River Basin. Pressurized air forced a small connecting channel through the seam. The coal around one hole was ignited and air, forced down the second, maintained the combustion reaction which produced carbon monoxide, hydrogen and methane. The mixture of these gases was piped to the surface for use as fuel.

The Department of Energy (DOE) is now supporting half a dozen development efforts, but none of these have yet reached the pilot plant stage. At the present rate of progress it may be 1990 before any full scale plant is operating. Eric Reichl, a respected coal scientist, is reported to have said, "Technology is not the key to success in coal conversion; politics is."[9] He asserts that "We could start building coal-to-gas plants right now. If we do not, this is again a political, not a technical decision. The issue after all is cost. Gas from coal will cost more, much more, than the price we have become accustomed to paying for natural gas." But not much more than we are paying for imported natural gas; given the prospects for the future and given the present halting progress, the crossover point will come long before we are ready.

Even now, in a moderately cold winter, we can stretch out our supplies of natural gas only by cutting off large users. Look back at Chapter 4. The diminishing supply can only mean deeper cuts in the future, in spite of conversions to coal. By 1990, it may be already far too late.

---

[9] Ibid.

During World War II, most of the fuel that powered the Luftwaffe and the Panzer divisions was derived from coal. With little oil available, there was no alternative. Coal was first gasified, then using the Fischer-Tropsch process, converted to a hydrocarbon liquid fuel. It was a large-scale process and it obviously worked. The plants were prime targets for the Allied bombing and those remaining were dismantled at the end of the war.

The conversion of coal to liquid fuel with present methods is a more demanding technology than gasification; it involves not only taking apart the carbon-carbon chemical bonds in coal, but also putting them back together in new combinations. However, the science is in its infancy. New methods of liquefying coal directly offer great technical promise, but as yet little achievement. A variant of the Fischer-Tropsch process can convert coal to alcohol, primarily methyl alcohol, methanol, which as a fuel burns cleanly. It can be used as a motor fuel, either alone or in combination with gasoline. How much would it cost? Methanol and the new liquifying techniques being considered by D.O.E. would produce clean fuel at the equivalent of some $20 a barrel—that same figure again.

The only operating plant that exists in the world today to convert coal to liquid fuels is in Sasolburg, South Africa. Using the Fischer-Tropsch process and sited at the mouth of a coal mine, it has been in production for over twenty years. South Africa, with cheap coal, politically isolated and concerned about its imports of oil, has found the process worthwhile. Plans have recently been announced for the construction of a second and much bigger plant that will produce the equivalent of ten million barrels per year.

Our own effort to begin the production of synthetic gasoline of any kind has been almost nonexistent. Why? Who holds it back? Even research and development work is halting and tentative. The modern catalysts developed for petroleum refining make liquefying processes far more attractive than those used by Germany in the 1930's and 1940's. In spite of this, the Federal government's difficulties in obtaining cooperative research from the oil companies seem to arise from the disinclination of the companies to believe that the price of crude oil will rise to $20 or $25 a barrel within the foreseeable future. After all, imports are still available. But pilot plants must be built to resolve the host of engineering problems that should be anticipated, and production plants may be years behind that. To quote Allen Hammond[10] again: "In the United States . . . the political climate is decidedly unfavorable to a major subsidized effort to produce synthetic oil, and the *de facto* policy is to buy oil abroad and damn the consequences. But it would be unfortunate if political and economic uncertainties combine to delay development of the liquefaction technologies that the country will surely need, possibly sooner than later."

Need more be said?

Survival in the unwanted revolution depends on a massive switch from natural gas and oil together with the rapid production of manufactured fuel from coal. But there is an ironic footnote. Even this may be only an interim solution. Burning any of the chemical fossil fuels releases carbon dioxide, CO_2, to the atmosphere, and since the beginning of the industrial revolution the amount of CO_2 in the atmosphere has increased by 13 percent. There are persistent worries that increasing the level of CO_2 in the atmosphere will lead to global warming. A panel of the National Academy of Sciences on Energy and Climate indicates that if fossil fuels continue to supply the bulk of our energy, by the middle of the next century the level of carbon dioxide in the atmosphere will have doubled and the climate will be warmer, on average, by some 5F. This may shift the corn belt from Iowa to central Canada, where the soil is poorer. It would melt some of the Arctic and Antarctic ice and raise the sea level so that New Orleans would be facing the same battle of encroaching waters presently confronting Venice. As Roger Revelle, Chairman of the Academy panel said, "We will have to be prepared to go to other sources of energy than coal within a finite time, about fifty years. . . . We will have to kick the fossil fuel habit by 2050."[11]

[10] A. L. Hammond, *Science* 193, 3 September 1976, p. 873–75.
[11] *Science News* 112, no. 68, 30 July 1977.

Those of us who enjoy worrying can keep this in mind. For most of us, there are more immediate concerns. We must kick the oil and natural gas habit first.

Ninety miles northeast of San Francisco, in the Big Sulphur Valley, plumes of natural steam were thundering skyward before men were there to see them. Only recently has this energy from the bowels of the earth been harnessed—the Pacific Gas and Electric Company's geothermal plant "The Geysers" now generates about nine hundred megawatts of electric power to be fed into the grid. This is the equivalent of one good-sized coal fired plant—what are the prospects that the success here can be multiplied many times over until geothermal energy provides a significant contribution to our total energy supply?

Not, alas, very good. True, the interior of the earth is very hot and if this heat could be harnessed it would represent a virtually inexhaustible resource. The problem is that, in most places, one must drill to a depth of about eight miles before the temperature rises to 300C. With present drilling technology it is prohibitively expensive to drill beyond about four miles. In addition, heat flows through solid rock very slowly, so that if heat energy were to be extracted from such a well, the surrounding rock would cool very quickly. The practicalities of geothermal power require a location that is, in effect, a geological freak, a place where the high temperature extends close to the surface and where the rock is sufficiently fractured that natural (or injected) water can carry the heat upwards. Such locations are rare. There are some additional possible sites in California, but still insufficiently explored. In Italy, a geothermal power plant has been in operation since 1904 and presently produces about 360 megawatts, approximately one-third the output of the Geysers plant. Others have been built at Wairakei in New Zealand, in Mexico, Iceland, several areas of Japan, and in the U.S.S.R. The harnessing of geothermal power is certainly practical in places where it is available; unfortunately, there are not enough such places. Geothermal power is megapower, but only in scattered locations—a useful local supplement. It is certainly worth developing, because every little bit helps, but equally certainly it is not an overall solution to our national energy needs.

The same is true of tidal power. In the Bay of Fundy, the tidal range is about fifty feet, larger than anywhere else in the world. It has been widely speculated that a barrage across the mouth of the bay could trap the water at high tide, to be used to generate electricity as in a hydroelectric plant. However, the performance of such a device is likely to be very disappointing. The high tides in the Bay of Fundy depend on a peculiar type of resonance; not a great deal of tidal energy flows in from the open ocean, but once in, it is trapped and the size of the oscillations builds up. The idea of resonance is familiar in connection with sound—bathroom singers usually notice that certain tones reverberate in resonance—but the effect is not limited to sound waves. Even tidal oscillations, with frequencies of one or two cycles per day, can resonate, and the length and depth of the Bay of Fundy happen to be just right to produce resonance with the tidal oscillations. This is the reason that the tides are so high. If we were to build a barrier across the mouth of the bay, we would unfortunately destroy the resonance, and the remarkable tides would disappear. The result would have been only the destruction of a unique natural phenomenon.

Nevertheless, useful energy can be extracted from tides in a few locations. There is one medium-sized (five hundred megawatt) installation on the Rance River estuary in Brittany. If we harnessed all the tidal power available in the United States, it may again provide useful local supplements, but less than one percent of our national electrical energy needs.

One thing that nuclear power plants *do not do* is release carbon dioxide to the atmosphere—in fact, their atmospheric emissions are relatively clean. Yet it is difficult to recall any similar development program that has aroused such advocacy, such exaggerations, such passionate opposition, and such apprehension as has the nuclear power plant program.

The basic facts are fairly simple; some may be surprising to American readers. On the other hand, the web of interpretations, claims, and fears expressed by dozens of different groups has produced a mess so impenetrable that few of us have the perseverance to untangle it. Let us stick to the basic facts.

So far, the United States nuclear power plant program has an enviable safety record. During the last 25 years, nobody has been killed in a nuclear accident at any commercial power plant. During the same time more than 6,500 men have been killed mining coal. Nevertheless, the fear is that if there should be even one substantial nuclear accident, a great many people may be killed or injured by the radioactive debris.

Nuclear power plants generate electricity in the same way as do coal-fired plants—high-pressure steam passes through turbines connected to electric generators which produce the current. The only difference, and it is a big one, is that the furnace in a coal plant is replaced by a nuclear reactor, whose sole function is to provide heat. Fresh fuel for American power reactors contain about 3 percent of the fissionable isotope uranium 235, and the break-up of these atoms provides the heat. The rest of the uranium in the fuel (and in the natural ores) is the non-fissionable type uranium 238. By the time the content of uranium 235 has decreased to about 1 percent as the result of the nuclear reactions, the combined effects of depletion and the accumulation of poisonous bi-products (including plutonium 239) make it necessary to replace the fuel rods.

What can be done with the spent fuel rods? They can be reprocessed to separate the unused uranium 235 and the plutonium from the other radioactive products which are then stored away to decay naturally. The uranium 235 can be re-cycled. Plutonium is fissionable and it also can be used as a nuclear fuel. Unfortunately, plutonium can be used in a quite different way, to make atomic bombs, and that is one of the problems. Another problem is the fact that some of the poisonous wastes take a great deal of time to decay, so that several centuries of secure leak-proof storage are needed.[12] Can we be confident that any storage method will be leak-proof for several centuries? The penalty of failure is grave.

Nevertheless, much of the nuclear material *can* be re-cycled, but what is in fact done with our nuclear wastes? Very little. The first American civilian nuclear power station went into service at Shippingport, Pennsylvania in 1957, yet the United States has no commercial facility licensed to recover unburned uranium 235 and plutonium from spent fuel. Only one private plant was ever licensed to operate, a small one which was shut down in 1972 for modification and enlargement. Its owners, Nuclear Fuel Services, Inc., have since withdrawn their application for a license to re-open. At Barnwell, South Carolina, a plant with a capacity of 1,500 tons per year was completed in 1976, but is not in operation. *There is no recycling.* American nuclear power plants are operating on a once-through basis and the wastes are accumulating. Reliable figures are difficult to obtain, but presently there seem to be about 2,500 tons in storage and a further 1,000 tons are being added each year.

The situation in Europe is very different. A French plant on the Rhone at Marcoule has been operating since 1958 with a capacity of 1,000 tons per year and a second plant with similar capacity went into operation at The Hague in 1967. A consortium of British, French and West German enterprises anticipates an integrated re-processing operation by the early 1980's, capable of handling as much as 20,000 tons per year.

Back in the United States, our present once-through use of nuclear fuel cannot continue much longer. Look at the Hubbert cycles shown near the end of Chapter 4. Remember at the same time that less than 1 percent of naturally occurring uranium is the fissionable kind U-235. One way to break the grip of the Hubbert cycle would be to convert the non-fissionable U-238 to a usable nuclear fuel and this indeed is what happens in that most controversial of devices, the breeder reactor. The controversy arises because the usable nuclear fuel generated is again plutonium, which, when purified, is not only a

---

[12] A more detailed and clear discussion of the reprocessing of nuclear wastes is given by W. P. Bebbington, *Scientific American*, December, 1976. See also Harald Schutz, *Maryland Academy of Sciences Technical Notes* 27 and 29 (1977).

nuclear fuel but also a nuclear explosive. The prospect of a proliferation of breeder reactors throughout the world, all generating plutonium, is to some a nightmare that no energy shortage can dispel.

The United States, which pioneered the breeder reactor in 1951, has been uncertain whether to go ahead with it. Others are not so squeamish. In Britain, France, and the Soviet Union, commercial sized prototypes are already on line, and a small one in Japan was completed in January, 1977. The British plant at Dounreay, Scotland, works well but has had problems (with leaks in the plumbing!). The French Phénix reactor at Marcoule which uses liquid sodium as a coolant, is much more dangerous than conventional U.S. commercial reactors, but so far, it has been so successful that Superphénix, with five times the capacity, is now in preparation west of Lyons. Even if we are not building breeder reactors, others are.

So here in its essence is the dilemma. Our present use of nuclear fuel gives us a lifetime of a decade or two. There are few prospects here for the massive alternative energy sources that we need. Breeder reactors not only generate power themselves, they supply fuel for other "conventional" reactors. France, without oil and with little coal, has an energy problem much worse than ours. She plans to generate about 35 percent of her total energy supply from nuclear plants by the year 2000, and most of the nuclear fuel is expected to come from breeder reactors.[13] They could multiply our own uranium resources many times, but they could also put the capacity for making nuclear bombs into the hands of every country on earth. They offer the shining prospect of abundant power, but the price may be death.

The development of controlled nuclear fusion for power generation has been a dream for twenty-five years. Nuclear fission, the process used in present reactors, involves the disintegration of heavy

---

[13] *New York Times*, 8 July 1976.

radioactive atoms. Fusion is a quite different process in which light atoms such as hydrogen are put together to form new stable atoms that are not radioactive. Hydrogen, a major constituent of ordinary water, is in abundant supply—we need not worry about the Hubbert cycle here. Nuclear fusion is immeasurably cleaner than fission; there is not the awesome spectre presented by the disposal of radioactive wastes. Fusion has powered the sun for billions of years and will continue to do so when our species is extinct. Fusion is true megapower—a burst of it is enough to produce a hydrogen bomb.

The problem with fusion is, in essence, the problem of reproducing a part of the sun, of controlling it, and confining it at temperatures that would vaporize any material known. Twenty-five years of research has been devoted to this aim. Despite considerable effort and ingenuity, the problems remain unsolved—the goal of producing a continuous stream of energy from fusion continues to elude us. It is not that we cannot make it work economically, or even work reliably —we cannot make it work at all. Perhaps some breakthroughs will come to cap those twenty-five years of research and the dream of controlled fusion will become a reality, but on the other hand, perhaps the whole thing is intrinsically impossible. Perhaps the only way continuous fusion can be maintained is in a huge ball of gas hanging in space—in the sun itself or in the stars. Perhaps not, but in either event, can we as a people rely for our critical energy supplies on scientific breakthroughs not yet made, and which, indeed, may never come?

In summary, the immediate course for national survival in this unwanted revolution can only rest on conservation, both individual and national, and on a rapid development of alternative fuels based on coal. The ever-more frantic search for new deposits of oil and natural gas, advocated by industry lobbyists, can only accelerate the depletion of the limited resources remaining. Nuclear fission contributes only a small part of our total energy supply and it could still unleash a Pandora's box upon us and our children. Fusion does not yet work.

Conservation and coal conversion—there seem to be no immediate workable alternatives. The one is being forced upon us by the relentless increases in the prices of oil that lie ahead. The other can be done, it *has* been done, but will it be done again on the massive scale that is needed if our society is not to choke of economic strangulation? We are woefully unprepared.

8
Finding the Missing Megapower: New Sources

Just as there are two quite distinct scales in energy production and consumption, megapower and micropower, so also are there two distinct scales in the financing needed. The generation of micropower by domestic solar collectors represents a substantial personal investment; the development of new sources of megapower will require megamoney—investments that are huge even on a national scale.

Massive amounts of money are already involved in our present energy supply. Even a large and successful company like Houston Oil and Minerals has provided, as mentioned in Chapter 4, a miniscule contribution to our overall consumption. The Exxon Corporation in

1977 had a net worth of $20 billion,[1] and sales of over $53 billion; Exxon is just one of the energy giants. The domestic petroleum industry as a whole had a net worth in 1977 of $128 billion, the equivalent of well over $1,000 for every breadwinner in the United States. This is the kind of investment that is needed if we are to achieve even the interim steps such as the extraction of liquid and gaseous fuels from coal, and needed again for the large-scale development of new energy sources that must supplant them. The magnitude of the task is daunting. Financing new megapower will strain our national budget and investment capability to the limit, but the longer the commitment is postponed, the closer we will be to the end of our present resources and the greater the ultimate trauma.

Where are the new energy sources and how can they be tapped? We hear of many fanciful schemes. It is easy to dream of revolutionary scientific advances that will solve our problem and fun to fantasize about novel and daring ways in which they may be used. But the problem is upon us—we do not have the leisure to indulge in dreams. We have a present problem; we cannot rely on scientific breakthroughs not yet made. We can hope for them, we should seek them and work desperately to bring them to fruition, but we cannot *rely* on them either this year or next year or ten years hence. However, we do have an advanced and powerful technology and the trained minds that have built it. Solutions will not be found by retreating from technology but by using it in new ways, with ingenuity and awareness of the resources of this planet and with sensitivity to the environment about us. The technology we can apply is what we have now. The pertinent question is: what can be done with *present technology?*

Solar energy—the response is hackneyed, but the question is *how?* The basic problem with solar power is that the radiant energy from the sun is diffuse. To capture even supplementary micropower for domestic use requires that the roof of a house be virtually covered with solar panels. To provide megapower for heavy industry or to run a whole city, the solar collectors must cover many square miles. If, indeed, solar energy were not so diffuse *we* would not be here—if its intensity were even as high as that which bathes the planet Mercury,

---

[1] Exxon, *Value Line Investment Survey*, January 20, 1978.

this earth would be uninhabitable. The sun's rays are benign; they are mild enough to support life but not to extinguish it.

What happens to the energy streaming constantly to us from the sun? Most of it, in fact, radiates back into space and is lost forever. On a clear night in winter, the temperature drops rapidly as the heat from the ground and the atmosphere radiates upwards. A small fraction of the energy, heating the atmosphere in places while it cools in others, is converted to winds which, in turn, drive the great circulations of the ocean and raise waves upon the sea surface. Some of the energy is used to evaporate water from the sea, to be released again in condensation and precipitation. A tiny part, about one percent, is used by plants in photosynthesis, yet this maintains all living matter in the world today. A miniscule part of this tiny fraction gave us the fossil fuel beds—the coal, the oil, and the natural gas—but it took aeons of time to compensate for the diffuseness of solar energy, the inefficiencies of conversion, the losses and the hazards of preservation. These resources are truly non-renewable.

The sun, then, is the primary source for most of our energy resources. Those that are renewable depend not on exploiting the accumulated legacy of past ages, but on capturing a part of the natural energy flow that is constantly streaming through the earth's system. Unfortunately, the fraction of the incident solar energy that is converted to other natural energy sources, such as wind or growing plants, is extremely small. If we were to regard it merely as a heat engine, the atmosphere is extraordinarily inefficient—only one four hundredth of the solar energy intercepted by the earth is converted to energy of motion in the winds. The fraction of all the wind's energy that we could conceivably intercept is miniscule; most of it is dissipated in the eddies and the turbulence as wind blows over the earth's surface. Only a tiny fraction of the wind energy is transmitted through the surface of the sea to produce the great ocean waves and systems of currents such as the Gulf Stream.

Since the natural energy conversion processes are usually so inefficient, the closer we are to the source the better off we usually are in seeking to divert a part of the energy flow for our own use. The direct conversion of solar energy to electricity is a more efficient process (if it were economical) than capturing second-hand energy from the winds. Even a very crude solar collector is a more efficient energy conversion device than the atmosphere. Nevertheless, there may be some circumstances that help to compensate.

Energy in ocean waves is third-hand solar energy. The winds, blowing over the ocean surface, raise waves but again only a small fraction of the energy flow driving the winds ends up in the form of ocean waves. Yet the ocean partially makes up for the inefficiency of conversion by its size—the waves that thunder on the west coast of Scotland have been generated in storm systems across the whole North Atlantic. As the great storm swells grow and propagate, they

carry the energy with them. It is collected from the wind over a large area, propagates to the shoreline where it is dissipated as the waves break upon the rocky coast. Cannot this energy flow be intercepted before it is lost in the foaming turbulence along the shore?

Such has been a dream for several generations, but only recently have devices been designed to achieve this end that are even remotely practical. One type, called the Salter Duck after its British designer, S. H. Salter, consists of an articulated string of curiously shaped floating tanks, like a floating breakwater. In the laboratory, these have achieved a remarkably high collection efficiency of the wave power. As waves impinge upon them, their nodding motion (hence the name "duck") is converted to electrical energy by an internal generator. Some have proposed the construction of strings of Salter Ducks some little distance off the western Scottish coast, where the water is shallow enough for mooring and the waves are chronically rough. They could indeed provide a significant contribution to Britain's electrical energy needs. It is an insteresting dream . . . but, as always, there is a catch! Practicalities will intrude, problems that are prosaic but may be insuperable. The overriding difficulty here may be the apparently simple task of anchoring the string in place. In fairly large waves, the average wave thrust on the Salter Ducks can be as large as 50 tons per yard; if the string is anchored every hundred yards, the anchors must hold and their chains resist a force of five thousand tons. Not even the largest ships have anchors that can support this!

One may be a little skeptical about whether strings of Salter Ducks will ever be built, but they do offer an instance in which even third-hand solar power might be harnessed. They may or may not be feasible, but there are more direct ways of capturing solar energy.

Radiation from the sun can be converted directly to electric current by means of photovoltaic silicon cells. These high technology products of the electronic age already operate photographic exposure meters, light meters, and automatic exposure cameras. Arrayed in panels, they already provide power for satellites in space. They have no moving parts, they are quiet, reliable and environmentally benign. They bypass the losses that are inevitably involved in the conversion of heat to electricity in conventional generating plants. They are also inordinately expensive in terms of the power they supply.

A single electric toaster requires about one kilowatt of power. An array of silicon cells to provide this power in full sunlight would presently cost some $15,000. The generation of megapower in this way with present technology would require an investment that is out of the question.

In the early years of the space program, silicon cells cost even more than they do today, but if they are to compete economically with the alternatives, the cost of production must be reduced to at most one-thirtieth of the present level. This is much more than can be anticipated simply from the economies of mass production. A target of $500 per kilowatt of generating capacity has been set for 1986 by the Environmental Research and Development Agency (ERDA), though it is not clear how this can be achieved. Here is an area in which one of those breakthroughs is needed but, ironically, according to Allen L. Hammond, ERDA is "casting photovoltaics as strictly a long-term option and is severely restricting its funding, [appearing] intent on ignoring both the stated objectives of its own sub-program and the signs of dynamism in private industry."[2] The technology in this area *is* expanding rapidly. Ideas for potentially economic processes to concentrate the energy and to manufacture the cells abound. It is not inconceivable that private industry can essentially "go it alone," and if the effort is successful, it will represent one of the most significant achievements of this century. The stakes are high and the need is urgent, but as far as ERDA is concerned, the gamble is worth less than one twentieth of what is spent to build one nuclear submarine.

Other solar energy collectors and converters are cruder than photovoltaic cells, but they are generally much cheaper, and they work. Most of them work only when the sun is shining, and this either restricts their usefulness to those energy demands that can be satisfied during the day, or else poses problems of energy storage. Moreover, solar energy by its very nature is dispersed—in a given geographical location, the sun shines more or less uniformly over the whole area. If

---

[2] *Science* 195, 29 July 1977, p. 447.

we are to convert solar energy to megapower, it must be collected over a large area and then concentrated by one means or another.

But does this really make sense? There is no doubt that in the generation of electric power by either coal-fired or nuclear plants, or in the production of synthetic liquid or gaseous fuels from coal, there are economies of scale. These enterprises require large and expensive plants for the conversion of energy to the form that we need, and, as a result, they also require elaborate and extensive distribution networks to convey the energy to where it is needed. In the search for missing megapower, it is perhaps natural to try to force solar energy production into the same mold, but to do so ignores the most natural and desirable qualities of this energy source and also its shortcomings. Solar energy is dispersed—why not take advantage of this to capture it where it is needed on the scale that it is needed? A large and expensive centralized solar energy installation is never going to provide the steady base-load electrical energy that we need night and day. Solar energy as we ordinarily conceive it is ill-suited to the generation of megapower, but it is perfectly suited to *relieve us of the need* for some of that megapower by the satisfaction of myriad, dispersed energy needs. We may not need to solve the difficult problems of large scale energy storage during the next twenty years, since even if massive developments are undertaken and the most optimistic projections are realized, during that time solar power will generally supplement rather than supplant our conventional energy sources. As Amory Lovins describes it, it is a soft technology, yet in 1977, over 60 percent of the federal research money devoted to the conversion of solar energy to heat and thence electricity was consumed in the effort to turn solar power into centralized megapower. It is an interesting effort, but may it not be fundamentally mis-directed? Rather than forcing solar energy at great expense into the almost incompatible mode of the established utilities, should we not take advantage of what it is and what it truly offers?

Simple flat plate collectors of the kind that will surely appear more and more commonly on the roofs of our houses can provide hot water at temperatures up to 100C, adequate for domestic use and for space heating. To run a heat engine for pumping water or to provide an on-site source of steam for industrial use, we need rather higher tem-

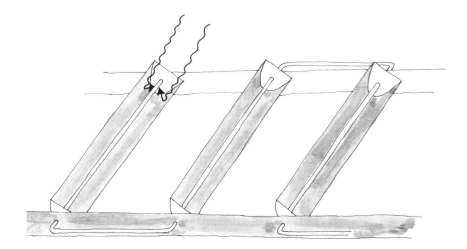

peratures; we need a device a little more elaborate than a simple flat plate. The *amount* of energy that we capture depends on the total area of the collectors; the *temperature* that we achieve depends on the extent to which that energy is focused. The sun's rays, focused by a magnifying glass, can ignite a piece of paper. Similarly, a collector in the form of a curved mirror can reflect the energy to a focus where it heats a liquid (water or freon) flowing along a tube. In order to keep the point of focus the same throughout the day, the curved mirrors, usually in the form of troughs, must rotate to track the sun. It is slightly more complicated but the technology for doing it is at hand—it has progressed rapidly in the past few years. By 1977, at least ten varieties of tracking collectors were being made in the United States, each in small numbers, but at costs comparable with simple flat plate collectors. Installation is more expensive. New types of non-tracking collectors, more effective than flat plates but with comparable cost, are appearing rapidly. One design consists of banks of evacuated glass tubes, like fluorescent lights, but containing a second, heat absorbing tube along the axis, through which flows the fluid to be heated. The vacuum between the two tubes gives excellent insulation against cooling by winds or cold outside air. Evacuated tube collectors are also more effective than flat plates in the morning and evening when the sun shines obliquely to the surface. Cost projections in 1978 for these and several other new types of collector range from about $10 per square foot with mass production. But mass production is slow in coming.

According to William D. Metz,[3] about 30 percent of all industrial process heat is used at temperatures below 300C, it is generally needed only during the day, and is well within the capability of tracking collectors to provide. Yet in 1977, ERDA, whose mission includes the exploration and demonstration of alternative energy sources, had only three projects in hand involving solar energy trough collectors. One of these was to provide hot water for washing cans at a Campbell Soup Company plant in California, another one to supply steam for fabric drying at an installation in Alabama, and a third one was to supplement the hot water supply of a concrete block plant in Pennsylvania. The imaginative leadership seems to be elsewhere.

In dry areas, the sun is often a more consistent and more dependable source of energy than the wind for purposes such as pumping irrigation water. The first (and still the largest) solar pumping installation was built, not under the auspices of ERDA, but by the Batelle Memorial Institute, a private research and development laboratory, with the support of the Northwestern Mutual Life Insurance Company. At Gila Bend, Arizona, it has worked very well with little maintenance. Similar systems could replace over three hundred thousand irrigation pumps in the western United States and with economies of mass production, the costs would be more than competitive, even if one ignores the future escalation of conventional energy costs. Here is an instance in which the way has been shown in a far-sighted private venture—we need many more of them!

A very small fraction of the tax dollars that we pay is used to stimulate the search for alternative energy sources—we each contribute only about $1.50 per year to solar energy research. Most of this is channeled toward the development of large and centralized energy systems which, as we have seen, may fail to exploit the most promising aspects of solar energy. Relatively small devices, such as that developed by Batelle and Northwestern Mutual, can be made as efficient as larger scale systems; they match the energy output to the

---

[3] *Science* 197, 12 August 1977, p. 651.

needs and provide it where it is needed. They accept solar energy for what it is. Yet as late as 1977, the thrust of the Federal Energy Research Program was in another direction.

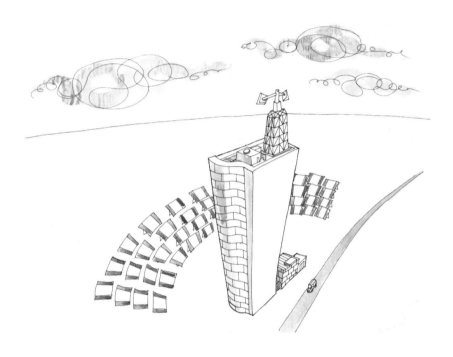

The present attempt to produce megapower from solar radiation centers on the "power tower" project. The idea is to collect the reflected energy from a field of mirrors, each of which tracks the sun, focusing it at high temperatures upon a single boiler mounted high in a tower. Steam from the boiler will provide electricity from a conventional turbogenerator. It is an interesting conception with a catchy name.

The first power tower in the United States, a pilot project, was built with ERDA's support at the cost of $21 million near Albuquerque,

New Mexico. Banks of almost three hundred reflectors spread over one hundred acres focus the sun's heat toward the boiler at the top of a two-hundred-foot tower. When the sun is shining, the thermal output in this pilot project is about five megawatts; despite the scale of the power tower, this is only about one-five hundredth the output of the furnaces in a large conventional electricity generating plant. It is obvious that power towers cannot be scaled up five hundred times—if their use is ever to become widespread, it will be in units whose output (during the day only) is much smaller than present day generating plants. Even so, the sheer physical size (and the cost) of projected future power towers is impressive. Construction of a second unit with an electrical output of ten megawatts was planned to begin in 1978 at Barstow, California. Its tower, when completed, will be as high as a fifty story buliding. Cost—about $130 million. Daytime output—sufficient for a town of about ten thousand people. Equivalent investment—$13,000 per person. These two endeavors are consuming the bulk of the federal funding for solar research and development, but it will not be until the 1990's that a prototype commercial plant is expected. The Government hopes to share the cost of later projects with the utilities that will use them, but the enthusiasm of the utility companies will surely depend on the eventual costs. It is not even certain that power towers will ever be economical. An optimistic estimate by Richard Caputo[4] of the Jet Propulsion Laboratory gives an eventual capital cost of $2,000 per kilowatt electrical power, one seventh the present cost of photovoltaic cells, but still four times the cost of electricity from a nuclear power plant.

The direct collection of radiant energy from the sun by power towers is clearly limited by the fact that they can work only during the day. In seeking to overcome this limitation, some have dreamed of solar energy collection by synchronous satellites, orbiting the earth at an altitude of about 22,300 miles and remaining stationary relative to the ground below. Such a satellite could be outside the shadow of the earth and in full sunlight during almost the whole year. One idea is to use photovoltaic cells to convert the radiant energy to electricity

---

[4] "Solar Thermal Electricity: Power Towers Dominate Research," *Science* 197, 22 July 1977, p. 353.

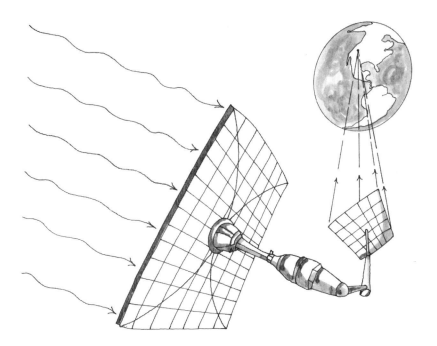

and thence to microwave energy which is beamed back to a large receiving antenna. Alternatively, a mirror could simply reflect the sun's energy to receivers and thence to a power tower, that would operate night and day. These wonderful fantasies are no more than that for several reasons. First, to collect megapower in this way, the area of the orbiting collection system must be of the same general size as those in a power tower or a solar farm on earth. Despite the shielding effect of the atmosphere, more than forty percent of the energy intercepted by the earth reaches the ground. A satellite collector may not be subject to the atmospheric loss but conversion and transmission losses would be even greater. The costs would be prohibitive; according to Harald Schutz,[5] in order for this to be considered economical, the cost of launching a satellite would have to be reduced to one fiftieth of what it is now and the costs of the solar

---

[5] Maryland Academy of Sciences, *Technical Note 26*, October 1976.

volatic cells to one thousandth. Finally, one must expect any manmade device to go wrong at times. What if the satellite guidance system fails and the high intensity energy beam drifts from its desired collection point, cutting a searing swath across the countryside and across our cities! This is the stuff of a disaster novel, not the basis of a sensible energy device.

The collection of solar energy on a large scale must indeed be an important component of our ultimate energy supply and the drive to develop power towers is a recognition of this fact. But as we have seen, the difficulties are enormous, and the prospects limited. Maybe power towers will make significant contributions, but possibly not. Solar energy may well be a large part of the answer to our problems, but this may not be the way to do it.

Let us stand back and view the whole problem in global perspective. About two-thirds of the earth's surface is covered by sea; two-thirds of the solar energy incident upon the earth ends up in the ocean. What happens to this energy?

Very little of the heat of the sun's rays is reflected by the ocean surface. Most of it penetrates some little depth into the water but is rapidly absorbed, so that by a depth of 100 feet or so, even in the clearest water, all of the infra-red radiation and virtually all of the visible light has been filtered out. The stream of energy from the sun warms the surface layers of the ocean to a temperature of 70F or 80F in tropical and sub-tropical zones; it is stirred by breaking waves at the surface and by the wind to a depth of a few hundred feet. At the same time, the wind blowing over the entire ocean basin generates the huge current systems such as the Gulf Stream in the Atlantic and the Kuroshio off Japan's Pacific coast. The warm surface water is carried by these great gyres towards the poles, cooling gradually (mostly by evaporation) along the way. By the time the water has reached the North Atlantic near Greenland or the Southern Ocean near Antarctica, the temperature is only a few degrees above the freezing point. As it chills, it becomes more dense, until it no longer lies at the surface but plunges to the bottom, some three or four miles down. It then gradually moves back towards the equator, still very cold, ultimately rising again to be incorporated by mixing into the warm, light surface layer. The cycle completes itself. The interesting and important fact about this circulation is that even in equatorial waters, it is only the surface layer that is warm; the underlying deep ocean water is only a few degrees above the freezing point. The surface is warm, the bottom is cold.

It is perhaps evident on a moment's consideration that a temperature difference is the basic thermodynamic requirement for any thermal energy conversion plant. In an ordinary electricity generating plant, heat is put into the steam at the temperature of the furnace, and after passage through the turbines, most of the heat is released in the cooling water, more or less at the ambient temperature outside. It is the flow of heat from hot to cold that provides the energy source, part of the flow being diverted and captured in the form of electricity. The large temperature differences in a conventional power plant represent high grade energy, but the relatively small differences in the ocean make it at best a low grade energy source. Nevertheless, the vastness of the ocean and the enormous magnitude of the natural energy flow may more than make up. Is it possible to generate ocean thermal power by taking advantage of the temperature difference between the surface and the deep ocean, by diverting a small fraction of the energy that is captured by this biggest of all solar collectors and degraded naturally by the large-scale mixing processes?

In fact it is. It has been done and it will be done again. It remains to be seen whether it can be done reliably and economically.

On October 22, 1930, M. Georges Claude, a famous French engineer, addressed the metropolitan section of the American Society of Mechanical Engineers in New York. He was sixty years old, not a particularly modest man, but he had little reason to be. He had already many achievements to his credit. He had developed new ways of liquefying air and oxygen for industrial use and had pioneered the industrial uses of the then new and rare gases, argon, neon, and helium. His technique for the production of ammonia at very high pressures made possible the large-scale manufacture of synthetic fertilizers. The neon tubes that he invented were a boon to the sign industry at least, if not to mankind as a whole. On this occasion, however, he was not talking about these things. His topic was "Power from the Tropical Seas." It was not a theoretical lecture—he was describing what he had done.[6] He had been the first to generate ocean thermal power.

He conceded that the *idea* was not first with him but with his "dear master and friend D'Arsonval as far back as 1882." He continued: "It is fortunate that I was not aware of this at the time I became interested, for, as you know, a beaten track has little allure for a good inventor and quite probably we should have abandoned the trail before discovering what interesting things it finally led to." He described his conception of the device. Under a vacuum, the warm surface water boils "not in a quiet, moderate way as perhaps might be supposed, but with violence, with a kind of explosion, every drop bringing its own provision of heat." Cold water from the ocean depths in a second chamber would cause the low pressure steam to condense, "this steam rushing through the pipe connecting the two vessels at a speed of 500 meters per second. If now in the path of this hyperhurricane flow of steam we place a turbine, this turbine will run and generate motive power."

The whole idea must have seemed crazy. He admitted that "we had hardly uttered our proposition before protestations and objections fell on our heads like hail . . . I underwent once more an experience common throughout my whole career, namely, that it is much more difficult to fight against man than against matter." Nevertheless, he persisted. In 1927 he chose a site for his pilot ocean thermal plant in Mantanzas Bay, Cuba, and construction proceeded during 1929. Claude did not have the resources to develop a floating power plant and the site chosen represented a compromise. His pipes to draw up

---

[6] The text was published in *Mechanical Engineering* 52, December 1930, p. 1039.

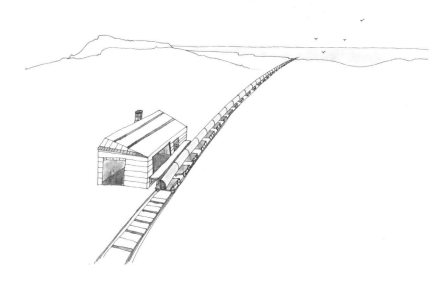

the cold, deep water were to lie on the sea floor; at the site chosen the bed was smooth enough but the depth available was regrettably small (only 2,000 feet or so) and the water temperature was not very cold (52F). The construction and laying of these pipes was his biggest task and it caused him endless trouble. On August 28, 1929, his pipes, a mile long, were floating at the mouth of a river five miles from the plant, ready to be towed into position and then submerged. But four of the tugs to do the job did not turn up, the weather worsened, the pipeline grounded and wrapped itself into an accordion. It was finally freed, but too late; the line was so damaged that before it could be installed, it sank in deep water, beyond reach.

Claude tried again. This time, he assembled his entire pipeline on trucks running along a track laid near the plant, from which it could be pulled directly into the sea. His new method of assembly almost worked. "Success seemed assured . . . when some of the men, acting stupidly and against their written orders, caused the head of the tube to sink with extreme rapidity, thus throwing a tremendous strain on the mooring cables, which gave way—and the second tube went to join the first."

Fortunately, he was not one to give up. On September 7, 1930, the third tube was successfully constructed, pulled to sea, submerged and attached. As he said to his New York audience, "Now, ladies and gentlemen, we see how right I was a moment ago in asserting that obstinacy is the best quality for an inventor to have, especially when it is backed up by a rational idea and a devoted personnel. Had I given way at the first or second failure, nothing would have remained of all this work but the souvenir of a foolish attempt." He added, without pretense of modesty, "As it turned out, however, all of my predictions came true and in the most natural manner."

Within a month of his success in attaching the pipeline, his plant was operating. Not with much efficiency, to be sure, since the temperature differences between his warm surface water and the cold cooling water was only about 22F. The power output was only 22 kilowatts, but the concept had been proven. Claude was conscious that great improvements could be made, even with the technology of 1930. He was optimistic that a plant to generate hundreds of megawatts could be built economically, and ended his speech with euphoria, claiming "that humanity has from now on the certainty that its industries will never lack the precious energy that actuates them."

But it was not to be, not in his lifetime at least. Shortly thereafter, a hurricane completely demolished his plant. Discouraged at last, he gave up. The 22 kilowatts that he achieved still stands as a record for ocean thermal power plants.

Claude's pioneering achievement was virtually forgotten for forty years. Only recently has come the realization that his exuberant prediction might after all be realized, that with present technology we may be able to achieve much better results than Claude ever did. The huge energy reservoir represented by the warm oceanic surface water is continually replenished by the sun; if only we could tap a small fraction of it before it is naturally degraded by oceanic mixing, we would have more than enough electrical energy for our needs, twenty-four hours a day. But how to do it? Will it operate effectively and economically?

By 1966, a few people, particularly the James H. Andersons, father and son, were beginning to explore the new potential for ocean thermal power made possible by the technological advances since 1930. Another pioneer is William Avery, a quiet, white-haired man not given to wild ideas, a senior scientist at the Applied Physics Laboratory of The Johns Hopkins University. By 1973 he had realized that the generation of significant amounts of power by the direct use of low pressure steam would require enormously cumbersome turbines, but if ammonia gas were used as the working fluid rather than steam, much higher pressures would be attained and much smaller turbines required. The ammonia gas would remain in a closed cycle. Under high pressure, it is a liquid, but heated by the warm ocean surface water, it boils. The high pressure gas then drives a turbine attached to an electric generator. As the vapor expands through the turbine, it cools. The exhaust gas is then further cooled and liquified by the very cold water pumped up from the ocean depths. The liquid ammonia returns to the evaporator to complete the cycle. The principle is precisely the same as Claude demonstrated; the substitution of high pressure ammonia for Claude's low pressure steam makes for a more compact, more feasible large-scale plant.

The Andersons and Avery are not the only advocates of ocean thermal energy conversion (OTEC). Groups at the Applied Physics Laboratory, at TRW (formerly Thompson-Ramo-Wooldridge), and Lockheed have devised various configurations of floating OTEC plants, drawing heavily upon the technology developed for oil drilling rigs. They visualize huge floating structures constructed, as are the large North Sea drilling rigs, from reinforced concrete but displacing as much water as a laden supertanker and generating 500 megawatts of electric power. Suspended beneath each of these giants of the sea would be an enormous tube, one hundred feet in diameter and 2000 feet long, to draw up the cold deep cooling water. It would dwarf any marine structure yet attempted. Nevertheless, these groups have examined again and again the potential costs of such plants, the problems of station keeping, the output of electrical energy and what to do with it. Large scale plants seem certainly within the grasp of present technology, and practical from the engineering point of view, at least as far as one can tell without a full-scale demonstration project. Nevertheless, considerable skepticism remains, with the "protestations and objections falling like hail" that Claude had experienced fifty years ago.

The skepticism is not unfounded. For one thing, the sheer size of OTEC plants is daunting. As a 1978 study by the U.S. Office of Technology Assessment notes, the cold water intake pipes would be "the equivalent of 20 or 30 Baltimore Harbor Tunnel tubes hanging vertically in the ocean." One of the most critical elements in the whole project is the efficiency of the huge units to transfer heat from water to ammonia, the heat exchangers: a thin layer of marine slime could so reduce their efficiency that the whole plant may become ineffective or even grind to a halt. How can we keep them clean? If an OTEC plant were forced to shut down for a few days, barnacles may begin to attach to the heat exchanger surfaces and remain there when operation resumed. How much would it cost to de-barnacle a whole system that had accidentally become encrusted? We simply do not know, nor are we likely to find out with certainty until the system has been tried.

Ocean thermal energy has been described as the biggest gamble in solar power.[7] If it comes off, it could provide abundant baseline power for many of our needs—it is the only solar system yet proposed that could do this. If it does not, it will have been an expensive failure, but still insignificant compared with our other national expenditures. The 1978 federal budget for OTEC is $35,000,000, about 15¢ for each person in the United States.

If it were to be successful, if we did develop a series of floating power plants scattered in the Gulf of Mexico, along the South Atlantic and Pacific coasts and near Hawaii, what would OTEC do for us? It is one thing to generate abundant amounts of electrical energy in the middle of the ocean, but quite another to deliver that energy to the places where it is needed. The present technology for transmitting large amounts of electrical power by undersea cables over great distances seems somewhat marginal; OTEC plants that feed directly into

---

[7] William D. Metz, *Science* 198, 14 October 1977, p. 178.

the transmission grid would necessarily be sited near shore. Yet deep water is required for the cold water source, so that the number of useful sites is limited.

On the other hand, this is not the only use to which the energy can be put. It can be converted to chemical form by electrolysis of sea water, producing hydrogen which, pressurized and liquefied, can be shipped back in tankers. Hydrogen is a perfect non-polluting fuel. It can be used in many applications to replace our dwindling supply of natural gas, and burns simply to water vapor. Alternatively, as Avery suggests, the electric power could be used on site for those industrial purposes that presently consume vast amounts of electrical energy.

One example is the production of ammonia itself for use in fertilizers. Ammonia can be synthesized from atmospheric nitrogen and from hydrogen produced by the electrolysis, together with abundant electric power. Another example is the refining of aluminum. At present, it is economical to dig aluminum ore, bauxite, in Tasmania, Australia, ship it to Zaire, where there is abundant hydroelectric energy, for processing by electrolysis, and then ship it on to the United States. A single 500 megawatt ocean thermal plant, used for the refining, could produce about 4 percent of the total U.S. production of aluminum, and, according to Avery, do it economically. A single such plant could satisfy our total needs for magnesium.

How much would it cost? The design studies suggest that an initial OTEC plant would cost from $1500 to $2500 per kilowatt, more than nuclear power plants, but much less than some of the alternative energy sources we have previously considered. It remains to be seen whether these estimates are realistic; on the one hand, final expenditures on new projects have an unfortunate habit of escalating, but on the other, the technology is not so esoteric, except for scale, that the estimates are blind.

How would these plants affect our oceanic environment? One clear consequence is that bringing to the surface layer such large quantities of cold water would reduce the water temperature at the intermediate depth where it is discharged. But remember that the amount of additional mixing is only a tiny fraction of that huge natural circulation that continues night and day, year after year. The temperature effect would be equivalent, in fact, to a shifting polewards of the internal oceanic 'climate' by only a few miles. As far as can be seen, the environmental consequences are extremely benign. There may be, in fact, enormous benefits. The deep ocean waters are nutrient-rich but sparsely inhabited, while the life in the sea, the plankton, the crustacea, and the fish dwell predominantly in the surface waters into which light penetrates. In isolated places of the world's oceans, the deep, nutrient-rich waters rise to the surface in a phenomenon called upwelling. It occurs along the coasts of Peru, Northern California (intermittently) and in various other places; the conjunction of the nutrient-rich waters and the photic (lighted) zone supports some of the richest fisheries on this planet. Occasionally, in El Niño, the natural upwelling off the coast of Peru disappears, and the fisheries suffer a precipitous and disasterous decline. Ocean thermal plants produce artificial upwelling on a smaller scale, to be sure, but with a number of such plants, the ultimate benefit to our food supply may possibly be as great as to our energy supply!

The present OTEC program is developing cautiously. It is to include three steps leading up to a commercial-sized demonstration plant, scheduled for installation in 1984. A gamble it is, and when one gambles, one must be prepared to lose. Nevertheless, the potential winnings are high relative to the investment. It can be true megapower, a renewable energy source on a huge scale whose envirionmental impact is minimal. At the very least, it represents the kind of imaginative and daring endeavor that can have a profound impact on our future energy needs.

BIOMASS CONVERSION SYSTEM

"Biomass conversion" is a presently fashionable jargon term that means, simply, energy from plants. The simplest biomass conversion system is a wood-burning stove.

This aspect of the energy scene is a curious combination of everyday experience, wild ideas, routine industrial economies, feeble action on the part of ERDA,[8] interesting private projects, and a dramatic drive towards accomplishment not here, but in Brazil. The domestic use of firewood in locations where it is available and in dwellings where wood-burning stoves or furnaces can be installed will surely become even more widespread. It is not so generally appreciated that already the wood-products industry, for example, derives about 40 percent of its total energy needs from burning bark and mill wastes,[9] nor that wood can be as easily gasified as coal. Sugar and grain crops, even when spoiled or of poor quality, can be fermented to produce ethyl alcohol, an excellent liquid fuel, while waste vegetable matter can be digested in the absence of air to produce methane gas. The range of biomass options then offer potentially renewable solid, liquid, and gaseous fuels that can in many instances replace directly those dwindling resources of oil and natural gas.

The basic questions are, as always, costs, but also in this case whether the large scale production of such fuels would compete seriously with land use for growing food and fibers such as cotton and, in other countries, jute. More immediate are the possibilities of using forest and agricultural wastes. For example, a test gasification unit, constructed by the California Energy Resources Conservation and Development Commission at a Diamond-Sunsweet walnut processing plant has so successfully used walnut shells (of all things!), that a larger gasifier is now being built by the company to provide virtually all of its energy needs at less than half the cost of the natural gas now being used.[10]

ERDA's initial efforts in this area were concentrated upon the extraction of energy from municipal trash. A pyrolysis plant has been working intermittently in Baltimore but even when operational, it provides more of a solution to the trash problem than it does to the energy problem. It takes more energy to produce the trash than one can obtain from it by pyrolysis. More recently, the attention of the

---

[8] And its successor, the Department of Energy (DOE).
[9] A. L. Hammond, "Photosynthetic Solar Energy: Rediscovering Biomass Fuels," *Science* 197, 19 August 1977, p. 745.
[10] Ibid.

Government Energy Agency has moved toward the prospects for effective production of liquid and gaseous fuels, but the investment is miniscule—less than $10 million per year, the equivalent of about 4¢ per person in the United States.

Nevertheless, lots of ideas abound, some possibly crazy, but some may be spectacularly successful. There is little real development, almost no accomplishment. The Batelle Memorial Institute sponsored a study that concluded that ethanol (industrial alcohol) *could* be produced in a full-scale plant from sugar cane or sweet sorghum at costs from $1.00 to $1.25 a gallon. In Brazil, it is already being done. With a properly tuned car, ethanol is superior to gasoline and even in the United States, the price differential is narrowing rapidly. But, on the other hand, sugar cane requires relatively good soils, which could otherwise be used for growing food; if forced to make the choice, we would rather eat than drive. Melvin Calvin, who won the Nobel Prize for his work on phytosynthesis, has shown that the gopher plant which grows wild on poor land in northern California, has a milk-like sap that is rich in hydrocarbons similar to those in petroleum. Perhaps we can grow our own gasoline! Calvin, working in Berkeley, is also seeking to develop a man-made photosynthesizer. In the leaves of plants, sunlight and some complicated chemistry breaks water down into hydrogen and oxygen; Calvin's device, if successful, could accomplish the same end with much greater efficiency. Water hyacinths which clog many inland waters and algae growing on sewage have been considered as sources for methane. These are only a few of the many ideas, but in this country there is still little serious attempt to bring them to fruition.

In striking contrast is Brazil's ambitious and determined national effort to grow its own fuel. Importing more than 80 percent of its oil, Brazil's economy was hit hard by the sharp rise in international oil prices. The country could not afford the luxury of protracted debate any more than we can, but in response to the problem, it did act decisively. In the last two years, over $400 million[11] has been committed to the task of producing alcohol as fuel on a sufficiently mas-

---

[11] A. L. Hammond, "Alcohol: A Brazilian Answer to the Energy Crisis," *Science* 195, 11 February 1977, p. 564.

sive scale to reduce the oil imports significantly within a few years; the program is developing momentum, and there is little doubt that it will play a large role in the country's future.

For years, Brazilian gasoline has contained up to 8 percent of ethyl alcohol, produced as a bi-product of sugar refining. No adjustments are required to a car engine burning this mixture and atmospheric pollution is reduced. The expansion in alcohol production now being undertaken will not, however, be based on sugar cane, since the rich land suitable for its cultivation is limited. An alternative source is manioc, a root crop, already widely grown as food. It is estimated that less than 2 percent of the land area of Brazil could produce enough fuel to replace all imported petroleum. New enzyme processes for converting the starch in the root to alcohol have been developed, and commercial production has begun.

It is still too early to assess whether or not the Brazilian enterprise will attain its ambitious goal, or even what the precise balance sheet will be with energy produced on the one hand, and energy consumed by fertilizers, cultivation, and production on the other. The United States is not Brazil; even if their program is spectacularly successful, it certainly cannot be translated immediately to U.S. conditions of soil, climate, and agricultural costs. Nevertheless, without such endeavors there can be no success. Its potential is enormous. We may easily find that our neighbor to the south offers in this respect a model, while we, with our far greater technological resources, limp lamely behind.

9
The Last Chance

After all this, where are we?

The undoubted facts are that the United States faces a continuing energy problem, that it is real, and that it will become much worse before it gets better. Our own production of oil and natural gas is dwindling inexorably, and in spite of conservation efforts, we are being forced to import an increasing proportion of these primary energy sources. The cost to the nation is ruinous and underlies the massive national deficits year after year, the erosion in value of the dollar abroad, and contributes to the stubbornly high rate of inflation at home.

For us individually and nationally the problems are great. We have the burden of replacing some 75 percent of our total energy supply, or doing without it.

Until abundant alternative energy sources have been developed, the noose of a restricted energy supply will tighten inexorably, but not necessarily day-by-day or even year-by-year. From time to time, we can expect temporary reprieves when we may perhaps wonder if the whole thing was really true. There will be regions of the country where, for a time, the oil available exceeds the demand, such as on the West Coast after the opening of the Alaskan pipeline. There will be occasional periods of over-production of oil in the Middle East as in 1977–78, with the accompanying and welcome stability in price. Successful widespread conservation efforts on our part will have the same effect, and herein lies a danger. A year or two with a plentiful supply of gasoline and price stability can allow the longer range problem to slip from mind. We can, for a time, forget the prospects that will haunt our children in young adulthood. We can relax our efforts and forget that the next crunch is likely to bite deeper than the last. But even a sinking ship rises a little to meet the crest of an ocean swell; that gives no reason to stop the pumps.

What would be the outcome of a continuation of present trends? History offers little comfort. In 1900, Britain stood at her zenith, prosperous and confident, buoyed by her investments abroad and her empire on which the sun never set. Two world wars stripped away those investments and the empire dissolved; left with an aging industrial base, only now is she enjoying a fleeting economic respite as the wave of North Sea oil passes by. Though the qualities of mind and spirit may reflect only slightly the fortunes of nations, the quality of life does so strongly. As late as 1953, long after World War II, food was still rationed in Britain, with one egg a week and half a pound of butter per person, because the country could not afford to import more. Portugal and Spain, which together once ruled the world, now offer few of their citizens those opportunities to develop their lives that we take for granted. There is no divine right among nations any more than there was among kings.

Yet the United States *does* have a broad base of other resources, of technical ingenuity and experience, and of scientific ability that can be its salvation. Science and technology earned an indifferent reputation in the early 1970's, in the wake of widespread concern about chlorinated hydrocarbons, the nuclear spectre and dangerous food additives. Nevertheless, an essential part of the solution of the long-term energy problem is the sensitive use of science and technology, on both a large and a small scale, with awareness of their impacts both on people's lives and on the environment that surrounds us. As we have seen, ideas abound. There is great potential. There are areas of rapid development, illustrated by the proliferation of new and more effective designs for solar collectors and by the successes in producing photovoltaic cells that are cheaper and cheaper to manufacture. There is still some distance to go, both scientifically in the uncovering of new natural processes and technologically, in the refinement of new designs. There is a *long way to go* in the development of mass production, but the rapid pace of achievement in these areas gives grounds for faith that this can ultimately be achieved.

Other areas with great potential, such as biomass conversion, present astonishing confusion. The apparent ineptitude of ERDA and its successor, the Department of Energy, in failing to offer clear direction and encouragement, together with the near absence of high technology industries which might "go it alone," have conspired to make progress in this area weak, halting, and sporadic. It need not be so—we have the example of Brazil to the south. The technology of most biomass conversion presently being used in this country is still at the level of the early 1800's—wood burning grates or Franklin stoves. In yet another area, the production of synthetic liquid and gaseous fuels from coal, the technology is there and constantly improving, but one looks around in vain for large scale plants under construction.

Our political response to the problems that we face has been feeble almost beyond credence. During virtually all of 1977 and 1978 we, and the world, were exposed to the unedifying spectacle of a Congress, paralyzed and unable to devise even a minimal energy policy which, at best, is only a beginning. The near collapse of the dollar on money markets abroad during that time reflected not only the succession of massive national deficits but also indicated the lack of confidence there that Congress would ever pull itself together to generate such a policy. Our representatives as a group have certainly not distinguished themselves in this task.

Why, a reasonable person might ask, has the political response been so inept and so inadequate? Our legislators can certainly make their own excuses, but three reasons seem to stand out. In the first place, although most of our representatives would concede that there is an energy problem, few appear to have realized the awful logic of its depth and inevitability. Few legislators have time to read and thoughtfully digest for themselves the rather technical reports and other writings that have given a warning. There are many matters pressing for a Congressman's attention, many hot issues of the day, many constituents to please. Immediate problems that can be solved quickly will receive attention first. The energy problem is a long problem, a hard and complex one, and few Congressmen have yet much appreciation of it.

Secondly, those members of Congress involved actively in the debate are subject to great lobbying by special interest groups, each with its own point of view and its own advantage to press for. Energy represents very big money, and big money has big lobbying power. It was pointed out in an earlier chapter that even a small contribution to our overall energy needs represents very large amounts of money indeed. The gas-producing companies naturally want to press for a price de-regulation schedule that is to their greatest advantage; they owe it to their stockholders to maximize profits. Utility lobbyists press for and environmental lobbyists oppose the construction of new nuclear power plants. The large oil companies have their lobbyists, as do the small ones, high technology firms lobby for a place in OTEC or the development of photovoltaics. Lobbying is a fact of political life in the United States (and probably everywhere). Is there any wonder that Congress, surrounded by great pressures from all directions, was unable to move in any?

The third reason is intrinsic to the American political system. There is, in effect, an election every two years, off-year elections alternating with full Presidential ones. Any member of the House of Representatives, once elected, must immediately start electioneering for the next time. He must help in the quick solution of relatively easy problems, he must take stands on popular issues so that there will be something to show before he again faces the voters. The energy problem is not a two-year problem nor a four-year one; it will certainly not be solved before the next election. Even with massive and successful effort, it will be twenty years before new energy sources are contributing a significant fraction of our national energy supply. Solid achievements are likely to be slow in coming, but the decisions must be made quickly so that progress is possible. A politically monolithic country, whose leaders face only pro-forma elections, can embark on a series of five-year plans which may or may not be successful, but they are certainly easier to develop. We are not about to change our political system to achieve this; the only real alternative is continued public awareness of the problem, translated into consistent pressure that it be faced in all its aspects—political, economic, societal, and technical. There is no single solution, no magic panacea. Progress will come through a net of intertwining new energy sources of different kinds to complement one another and to satisfy the diversity of our needs, but, as yet, little has been accomplished. There is so far to go and we have hardly begun the journey.

Time is running out. Already it is almost too late. Look back at Chapter 4 and project for yourself where our own production of oil and natural gas will be at the end of the century. Will we have been forced into an ever more hectic and intensive scramble, competing at ever higher prices, for the remaining resources of the Middle East? These present years are truly our last chance.

In all the matrix of possibilities, what is inevitable and what is not? There is no doubt that the energy problem will not go away. There is no doubt that the cost of energy will continue to rise substantially in real terms, even faster in terms of current dollars when inflation is taken into account. This will certainly encourage continuing attention to conservation, and may well impinge uncomfortably on a careless style of life. We may be somewhat poorer, at least by conventional measures of the standard of living. On the other hand, we may regain a national will and develop a new style of harmony with the resources of this planet that will show the way for the rest of the world. Not inevitable is the continued inaction and indecision that plagued the Congress and the country during the mid 1970's, but it will cease only by the will of the people.

Fine words, but what can the individual American do?

First, we must quite literally put our own houses in order. In Chapter 6 a number of simple ways were given to reduce our personal energy consumption and to save money; an ingenious reader can probably find many others. We must accept the fact that energy costs will continue to increase and it is in our own interests to try to deflect as much of the increase as possible.

Secondly, there is the time-honored remedy for dissatisfaction: write to your congressman! Tell him or her that you expect him to take a long hard look at the problem for our own and our children's sakes. Insist that he take a positive part in developing policies and legislation that will be the best for the country as a whole, even though some special interest groups may come off rather worse than others. Legislators are influenced by lobbyists, but they are very sensitive toward their voters. In this respect, the short interval between elections is an advantage—if we insist that action is taken based on a long term view, reminding them that they are accountable at the next election, action will soon be forthcoming. What difference will your letter make? It is true, of course, that a couple of letters will make very little difference but a couple of thousand represent many votes that can go one way or the other. Ask your representative what he has accomplished in contributing to the solution—perhaps he is part of the problem! Ask him in person, ask him in writing, ask him in a letter to your local newspaper. Our representatives are responsible to us. Give them hell and keep giving them hell!

Progress is never uniform. We must accept that some promising endeavors will turn out to be dead ends despite all our ingenuity and effort. In 1926 an experimental rotor ship was built by Anton Flettner in Germany, with huge vertical cylinders rising from the deck which could be turned about their axes with a small engine. The turning rotors could (and did) extract drive from the wind more effectively than square rigged sails; to that extent the ship was a success. But she could still be becalmed despite her engine to turn the rotors; she was slow, clumsy, and rather unseaworthy in heavy weather, so that no more were built. Other methods of propulsion were, on balance, better.

Some new methods of energy capture will reign for a time only to be supplanted by technological improvements in other areas. For thirty years, transatlantic steamships used paddles. Marine propellers were known and had been tested years earlier, but paddle wheels were better matched to the slowly turning marine engines of the 1840's. When higher speed steam engines with sufficient power were developed, paddle wheels began to disappear from the North Atlantic. The great transatlantic liners themselves have, in turn, yielded to jet airliners. In the same way, as our new energy web develops, by design or by default, the first steps will not be the final ones, nor the first designs the best. In solar collector technology, the first steps are being taken—but where are they in some of the other areas? It seems that only by collective insistence will they come.

At the other end of the scale, our individual consciousness of the extent and duration of the energy problem will be reflected in our actions in a myriad of situations:

The headmaster of a local school can install a secure bicycle rack to encourage the children to cycle rather than catching a bus or riding in a car pool.

An electrician re-wiring an old house can suggest to his client that fluorescent lights in the kitchen will be brighter and save money.

At birthday time, father can be given a handsome sweater rather than an electric toothbrush.

A young couple, seeking to rent a garden apartment, can ask probing questions about the extent of insulation and even insist on seeing for themselves.

If you love to be outside on the water, why not get rid of the power boat and take up sailing?

Taxpayers! Any of us can direct to our representative pointed questions on the temperatures to which municipal office buildings are heated in winter or cooled in summer.

For a man working at a shipyard, nothing tastes better than a cold beer at home after a hard day. He can get it in returnable bottles rather than disposable ones or aluminum cans.

Perhaps you have a few spare dollars to invest. Why not put them into a small capital-starved company with a good energy idea and the technical and marketing know-how to back it up? The company may fold (that is a risk you take) but it might possibly become another Xerox.

Each of these is a little thing, but each is a tiny step in the direction in which we must go. The energy problem is not just a technical problem for the scientists and engineers, not just a capital problem for the economists, not just a political problem for the legislators, it is our problem.

And so we come full circle. We have seen many things that can be done, some that are being done, and a great deal that is not being done. As was said in the beginning, this small book offers no detailed blueprint of solutions; these can come only by the devoted and continuous application of the talents and energies of the American people. If we are successful, we need not apologize to the generations that follow.

Index

A

Air conditioning, 68
Alaskan oil, 38, 39
Alcohol fuel from plants, 123
    production of, in Brazil, 125
Anderson, James H., 119
Anthracite, 15
Avery, William, 119

B

Biomass conversion, 123–26, 129
Bituminous coal, 14
Breeder reactors, 98, 99
Brown, Harrison, 53
Brown coal, 14

C

Calvin, Melvin, 125
Carbon dioxide, atmospheric, 94
Claude, M. Georges, 115–18
Coal, 11, 88
    conversion to, 89
    gasification of, 90–92
    liquefaction of, 93
    origins of, 13
    reserves of, 51
    use of, 89
Conservation, 7
    of energy, 55, 56, 60, 66–74, 86
Continental drift, 18
Continental shelf oil, 5
Cycles of discovery and production, 29–32

E

Echosounding in oil search, 16
Electrical heating, 64, 65
Electricity generation
    early, 25
    losses in, 65
Energy crisis
    and conservation, 55, 56, 60, 66–74
    and conversion, natural, 104
    and costs, 64, 66–68, 79–81
        in commuting, 84
    grades of, 60
    and losses in conversion, 65
    and strategies, 61
    in Tudor England, 22–23
    and usage, 56

F

Fireplaces, 70

G

Gasification of coal, 90–92
Geological Survey, U.S., 41
Geothermal power, 95

H

Heat pumps, 73
Houston Oil and Minerals, 48
Hubbert, M. King, 27–28, 43
Hubbert cycles, 29–32
    distortions of, 33
    statistical nature of, 33

I

Insulation in homes, 66, 67, 70

L

Lignite, 14
Liquefaction of coal, 93
Lovins, Amory, 61

M

Megapower, 57–62, 111, 123
    financing of, 102

Methane from plants, 124
Mexican oil, 82
Micropower, 57–62

N

Natural gas, 45
    prices of, 46, 91
    production of, U.S., 45, 46
    wells drilled for, 46
North Sea oil, 82
Nuclear fission, 97
    and fusion, 99
    and power, 10, 17, 51, 97–100
    reprocessing and, 97
    wastes, 98

O

Oceanic upwelling, 122–23
Ocean thermal energy conversion (OTEC), 115–22
Oil
    consumption of, by U.S., 4, 35, 40, 86
    discoveries of, U.S., 34, 39
    formation of, 14
    overseas payments for, 55
    production of, U.S., 35, 40
    reserves, U.S., 5
    reserves, world, 81
    resource estimation, 36
        errors involved in, 37
    supply, Exxon's projections of, 44, 48
    traps, 15
    ultimate recoverable, U.S., 36, 37, 41, 43
Ore concentrations, 18–21
OTEC, 115–22

P

Peat, 14
Photovoltaic cells, 106
Power towers, 111
Proven reserves, 31

R

Resource estimation, 36
    errors involved in, 37
Resources, renewable and nonrenewable, 11–21, 28, 104

S

Salt domes, 16
Sea coal, 23
Sea floor spreading, 20
Shale oil, 82
Soft technologies, 62
Solar collectors
    domestic, 74–77, 108
    passive, 77
    tracking, 109
Solar energy, 8–10, 74–77
    collected by the ocean, 114–15
    collected by satellites, 112
    as megapower, 108, 111–12
    and power tower, 111–12
    pumping, 110
Steam engines, 24
Stourbridge Lion, 24
Synthetic gas, 90–92
    and gasoline, 93, 95

T

Tidal power, 96
Transportation, 84, 85
Treasure ship analogy, 28

U

Uranium, 17
    discoveries, U.S., 49
    production, U.S., 49

W

Wave energy, 105
Wegener, Alfred, 18
Windmills, 61, 77

Z

Zapp, A. D., 42

THE JOHNS HOPKINS UNIVERSITY PRESS

This book was composed in Linotype Optima Text by the Maryland Linotype Composition Co., Inc., and Mergenthaler VIP Olive Bold display type by Dean's Composition, Inc., from a design by Charles West. It was printed on 55-lb. Antique Cream and bound by The Maple Press Company.

Library of Congress Cataloging in Publication Data
Phillips, Owen M.   1930–
  The last chance energy book.

  Includes index.
  1. Energy policy—United States. 2. Power resources—
United States. I. Title.
HD9502.U52P5   333.7   78–20511
ISBN 0–8018–2189–4